# Introduction to Manufacturing Processes

Levi O'Brein

Willford Press
www.willfordpress.com

# Preface

Manufacturing refers to the production of merchandise from raw materials using machines, human labor and via biological or chemical processing. It includes all processes that are vital in the production and integration of individual components of a final product. It plays a crucial role in the automobile and aircraft industry, household appliance and furniture production, etc. Any stage in the development of a finished product from its raw materials is termed as a manufacturing process. Manufacturing starts with product design and material specification. Manufacturing processes may include casting, imaging and coating, molding, forming, machining and joining, besides others. This book provides comprehensive insights into the field of manufacturing processes. It elucidates new techniques and their applications in a multidisciplinary approach. In this textbook, constant effort has been made to make the understanding of the difficult concepts of manufacturing as easy and informative as possible, for the readers.

A detailed account of the significant topics covered in this book is provided below:

Chapter 1- The production of goods for sale using machines, labor and processing is called manufacturing. Many manufacturing systems are important for executing manufacturing tasks. The diverse topics covered in this chapter such as factory system, biomanufacturing, manufacturing execution system, lean manufacturing, etc. strives to provide an extensive understanding of the varied aspects of manufacturing and manufacturing systems.

Chapter 2- Manufacturing process management is a set of techniques and methods that are used to guide the manufacturing of products. This chapter has been carefully written to provide an easy understanding of the varied facets of manufacturing process management, such as near net shape, metal casting, subtractive manufacturing, molding, metal forming, etc.

Chapter 3- Casting is a manufacturing process characterized by the design of complex shapes by pouring a liquid material into a hollow cavity, which is then allowed to solidify. The topics elaborated in this chapter on evaporative-pattern casting, permanent mold casting, die casting, continuous casting, centrifugal casting, etc. will help in providing a better perspective about casting processes and techniques.

Chapter 4- Molding is a manufacturing process concerned with the shaping of a liquid or a raw material by using a rigid frame called a matrix or mold. All the diverse techniques of molding such as blow molding transfer molding, injection molding, compression molding, rotational molding and thermoforming, have been carefully analyzed in this chapter.

Chapter 5- Forming processes are manufacturing processes which utilize different stresses like shear, tension, compression, etc. to cause plastic deformation in materials for producing desired shapes. This chapter aims to provide a comprehensive understanding of forming processes and includes vital topics such as rolling, extrusion, swaging, explosive forming, electroforming, hydroforming, etc.

Chapter 6- Welding is a sculptural or fabrication process that is used to join materials like thermoplastics or metals through the process of fusion. This chapter discusses in detail the different methods and processes of welding such as arc welding, oxy-fuel welding and cutting, shielded metal arc welding, gas tungsten arc welding, gas metal arc welding, flux-cored arc welding, etc. which are commonly used in manufacturing.

It gives me an immense pleasure to thank our entire team for their efforts. Finally in the end, I would like to thank my family and colleagues who have been a great source of inspiration and support.

**Levi O'Brein**

1

# Manufacturing and Manufacturing Systems

The production of goods for sale using machines, labor and processing is called manufacturing. Many manufacturing systems are important for executing manufacturing tasks. The diverse topics covered in this chapter such as factory system, biomanufacturing, manufacturing execution system, lean manufacturing, etc. strives to provide an extensive understanding of the varied aspects of manufacturing and manufacturing systems.

Manufacturing is the processing of raw materials into finished goods through the use of tools and processes. Manufacturing is a value-adding process allowing businesses to sell finished products at a premium over the value of the raw materials used.

Humans have historically sought ways to turn raw materials, such as ore, wood, and foodstuffs, into finished products, such as metal goods furniture and processed foods. By refining and processing this raw material into something more useful, individuals and businesses have added value. This added value increased the price of finished products, rendering manufacturing a profitable endeavor. People began to specialize in the skills required to manufacture goods while others provided funds to businesses to purchase tools and materials.

How products are manufactured has changed over time. The amount and type of labor required in manufacturing vary according to the type of product being produced. On one end of the spectrum, products are manufactured by hand or through the use of basic tools using more traditional processes. This type of manufacturing is associated with decorative art, textile or leather work, carpentry and some metal work. At the other end of the spectrum, mechanization is used to produce items on a more industrial scale. This type of manufacturing does not require as much manual manipulation of materials and is often associated with mass production.

The industrial process used to turn raw materials into products in high volumes emerged during the Industrial Revolution of the 19th century. Before this period, handmade products dominated the market. The development of steam engines and related technologies allowed companies to use machines in the manufacturing process reducing the number of personnel required to produce goods while also increasing the volume of goods that could be produced.

Mass production and assembly line manufacturing allowed companies to create parts that could be used interchangeably and allowing finished products to be made more readily by reducing the need for part customization. The use of mass production techniques in manufacturing was popularized by the Ford Motor Company in the early 20th century. Computers and precision electronic equipment have since allowed companies to pioneer high-tech manufacturing methods. Products made using these methods typically carry a higher price but also require more specialized labor and more expensive capital inputs.

The skills required to operate machines and develop the processes used in manufacturing have changed drastically over time. Many low skill manufacturing jobs have shifted from developed countries to developing countries because labor in developing countries tends to be less expensive. More skilled manufacturing, particularly of precision and high-end products, tends to be undertaken in developed economies. Technology has made manufacturing more efficient and employees more productive; therefore, although the volume and number of goods manufactured have increased, the numb er of workers required has declined.

Economists and government statisticians use various ratios when evaluating the role manufacturing plays in the economy. Manufacturing value added (MVA), for example, is an indicator that compares manufacturing output to the size of the overall economy. It is expressed as a percentage of GDP - Gross domestic product. The ISM Manufacturing Indexuses surveys of manufacturing firms to estimate employment, inventories and new orders and is an indicator of the health of the manufacturing sector.

## Manufacturing Business

A manufacturing business is any business that uses raw materials, parts and components to assemble finished goods. Manufacturing businesses often employ machines, robots, computers and humans to produce the merchandise and typically use an assembly line, which enables a product to be put together step by step, moving from one workstation to the next.

Manufacturing businesses can choose to sell their products directly to consumers, to other manufacturers, to distributors or to wholesalers.

### Difference between a Manufacturer and a Wholesaler

A manufacturer is responsible for assembling the final goods, such as laptops, fridges or watches, or producing parts and components that other manufacturers use to produce more complex products like cars or aircraft. A wholesaler, on the other hand, is a middleman between a manufacturer and a retailer (or end consumer). It buys products in bulk from the manufacturer and resells them in smaller quantities for a higher price.

### Types of Manufacturing Production

There are three main types of manufacturing production: make-to-stock, make-to-order, and make-to-assemble.

Make-to-stock is a traditional manufacturing strategy that relies on past sales data to forecast consumer demand and plan the production activity in advance. The drawback of this strategy is that it uses past data to predict future demand, which increases the likelihood of the forecasts being off, leaving the manufacturer with too much or not enough stock.

Make-to-order allows customers to order products that are customized and manufactured to their specifications. The manufacturing process begins only after the order is received, so the waiting time for the customer is longer, but the risk of excessive inventory is cut out.

Make-to-assemble is a strategy that relies on demand forecasts to stock the basic components of a product, but starts assembling them after the order is received. It's a hybrid of MTS and MTO approaches. Customers can customize the products and receive them quicker because the manufacturer has the basic components ready, but if orders don't come in, the manufacturer is stuck with a stock of unwanted parts.

All three types of manufacturing businesses have certain risks involved. Producing too many goods leads to financial losses as money is tied up in unwanted stock; producing too little means not meeting the demand, which can cause the customer to turn to competition and induce a drop in sales for the manufacturer.

To reduce the risks, any type of manufacturing business should focus on keeping production costs low, maintaining good quality control and investing in excellent sales management.

## Manufacturing System

The term manufacturing system refers to a collection or arrangement of operations and processes used to make a desired product or component. It includes the actual equipments for composing the processes and the arrangement of those processes. In a manufacturing system, if there is a change or disturbance in the system, the system should accommodate or adjust itself and continue to function efficiently. Normally the effect of disturbance must be counteracted by controllable inputs or the system itself. Figure gives the general definition for any manufacturing system.

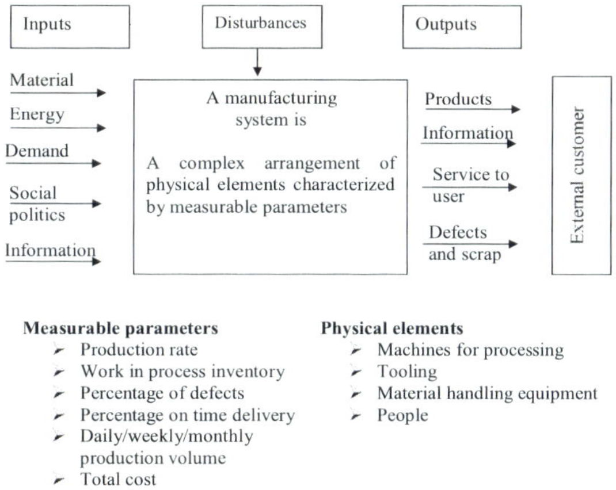


Figure: General definitions for any manufacturing system

## Types of Manufacturing Systems

The manufacturing systems differ in structure or physical arrangement. According to the physical arrangement, there are four kinds of classical manufacturing systems and two modern manufacturing systems that is rapidly gaining acceptance in industries.

The classical systems are:

1. Job shop,
2. Flow shop,
3. Project shop,

4. Continuous proces.

The modern manufacturing systems are:

1. Linked cell system (Cellular manufacturing system),
2. Flexible manufacturing system (FMS).

## Job shops

In a Job shop, varieties of products are manufactured in small lot sizes to a specific customer order. To perform a wide variety of manufacturing processes, general purpose production equipment is required. Workers must have relatively high skill levels to perform a range of different work arrangements.

The production machines are grouped according to the general type of manufacturing processes as shown in figure. The lathes are in one department, drill presses in another and so on. Each different part requiring its own sequence of operations can be routed through the various departments in the proper order. For this 'ROUTE SHEETS' are used. The layout made for this purpose is called as functional or process layout.

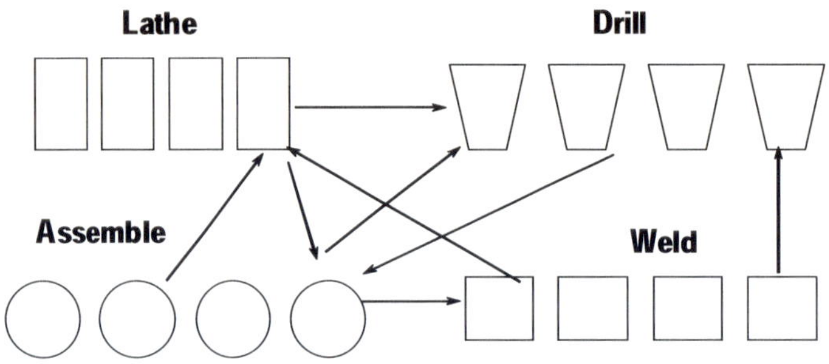


Figure: Functional or process layout

## Advantages of Process Layouts

- Can handle a variety of processing requirements,
- Not particularly vulnerable to equipment failures,
- Equipment used is less costly,
- Possible to use individual incentive plans.

## Disadvantages of Process Layouts

- In-process inventory costs can be high,
- Challenging routing and scheduling,
- Equipment utilization rates are low,
- Material handling is slow and inefficient,

- Complexities often reduce span of supervision,
- Special attention for each product or customer,
- Accounting and purchasing are more involved.

Examples: Machine shops, foundries, press working shops, plastic industries.

## Flow Shops

The flow shops have a "product oriented layout" composed mainly of flow lines. This system can have high production rates. The plant may be designed to produce the particular product or family, using "Special purpose machines" rather than general purpose equipments. The skill level of the laborer tends to be lower than in production job shop. When the volume of production becomes large, it is called "mass production". The material flow is through a sequence of operations by material handling devices. The time the item spends in each station or location is fixed and equal. The workstations are arranged in line according to the processing sequence needed as shown in figure.

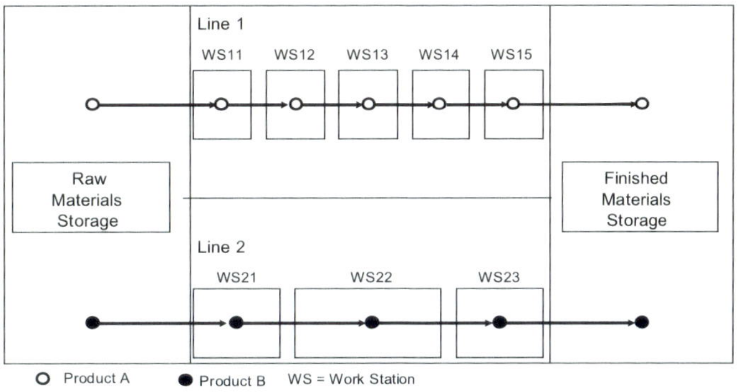


Figure: Product layout

## Advantages of Product Layout

- High rate of output,
- Low unit cost,
- Labor specialization,
- Low material handling cost,
- High utilization of labor and equipment,
- Established routing and scheduling,
- Routing accounting and purchasing.

## Disadvantages of Product Layout

- Creates dull, repetitive jobs,
- Poorly skilled workers may not maintain equipment or quality of output,

- Fairly inflexible to changes in volume,
- Highly susceptible to shutdowns,
- Needs preventive maintenance,
- Individual incentive plans are impractical.

Example: Automated assembly line and Television manufacturing factory.

## Project Shop

In this type, a product must remain in a fixed position or location because of its size and weight. The materials, machines and people in fabrication are brought to site. The layout is also called as fixed position layout. Figure shows the project shop layout.

Example: Locomotive manufacturing, large aircraft assembly and ship building

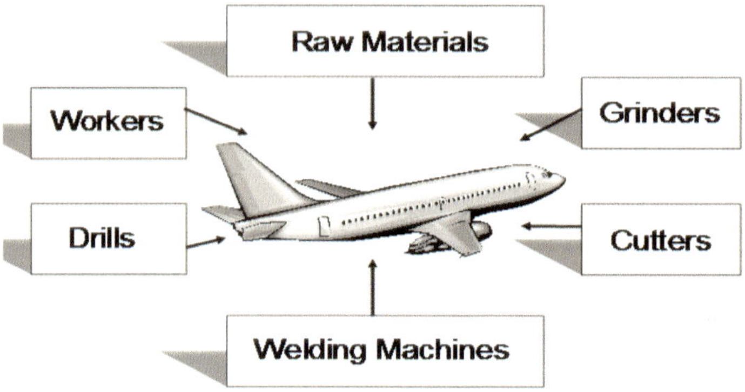


Figure: Project shop layout

## Advantages of Project Layout

- Minimum capital investment,
- Continuity of operation,
- Less total production cost,
- Offers greater flexibility,
- Allows the change in production design,
- Permits a plant to elevate the skill of its operators.

## Disadvantages of Project Layout

- Machines, tools and workers take more time to reach the fixed position.
- Highly skilled workers are required.
- Complicated jigs and fixtures (work holding device) may be required.

## Continuous Process

In this continuous process, the product seems to flow physically. This system is sometimes called as flow production when referring to the manufacture of either complex single parts, such as a canning operation, or assembled products such as TVs. However, this is not a continuous process, but high volume flow lines. In continuous process, the products really do flow because they are liquids, gases, or powers. Figure shows the continuous process layout. It is the most efficient but least flexible kind of manufacturing system. It usually has the leanest and simplest production system because this manufacturing system is the easiest to control because it has the least work-in-progress (WIP).

Examples: Oil refineries, chemical process plants and food processing industries

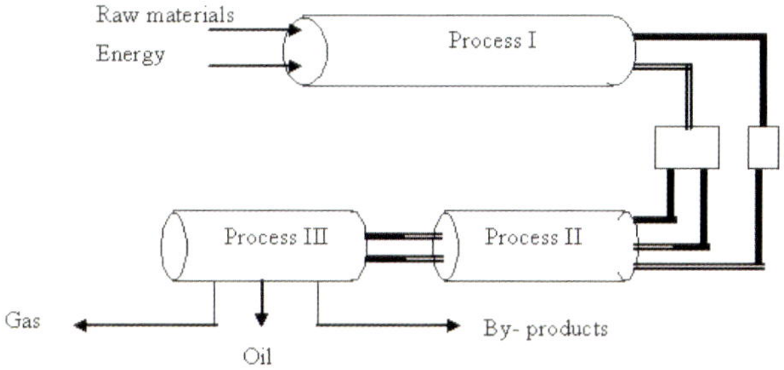


Figure: Continuous process layout

## Linked Cell Manufacturing System

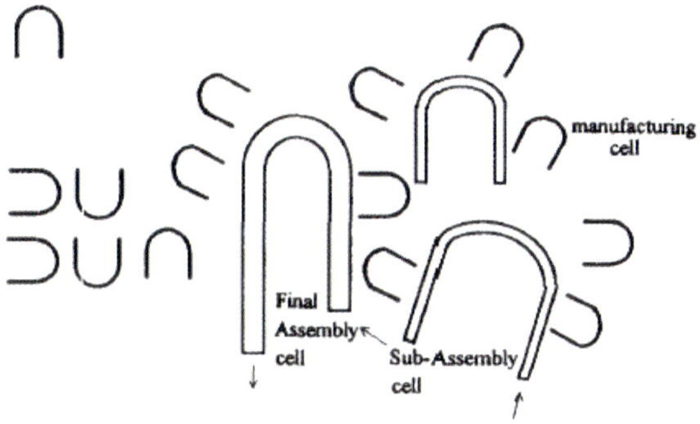


Figure: Main structure of cellular manufacturing System

Cellular manufacturing (CM) is a hybrid system for linking the advantages of both job shops (flexibility in producing a wide variety of products) and flow lines (efficient flow and high production rate). A cellular manufacturing system (CMS) is composed of "linked cells". Figure shows the main structure of cellular manufacturing system. In cells, the workstations are arranged like a flow shop. The machines can be modified, retooled and regrouped for different product lines within the same "family" of parts. This system has some degree of automatic control for loading and unloading of raw materials and work pieces, changing of tools, transferring of work pieces and tools between workstations. Cells are classified as manned and unmanned cells. In manned cells multifunctional operators can move from machine to machine and the materials can be moved by the operator. In the unmanned cells, an industrial robot is located centrally in the cell for material handling. Automated inspection and testing equipment can also be a part of this cell.

## Advantages of CMS

The advantages derived from CMS in comparison with traditional manufacturing systems in terms of system performance have been discussed in Farrington, Kannan, Suresh, Hug and Assad. These benefits have been established through simulation studies, analytical studies, surveys, and actual implementations.

They can be summarized as follows:

- Setup time is reduced: A manufacturing cell is designed to handle parts having similar shapes and relatively similar sizes. For this reason, many of the parts can employ the same or similar holding devices (fixtures). Generic fixtures for a part family can be developed so that time required for changing fixtures and tools is decreased.
- Lot sizes are reduced: Once setup times are greatly reduced in CM, small lots are possible and economical. Small lots also provide smooth production flow.
- Work-in-process (WIP) and finished goods inventories are reduced: With smaller lot sizes and reduced setup times, the amount of WIP can be reduced. The WIP can be reduced by 50% when the setup time is cut to half. In addition to the reduced setup times and WIP inventory, finished goods inventory is reduced. Instead of make-to-stock systems with parts either being run at long, fixed intervals or random intervals, the parts can be produced either JIT in small lots or at fixed, short intervals.
- Material handling costs and time are reduced: In CM, each part is processed completely within a single cell. Thus, part travel time and distance between cells is minimal.
- A reduction in flow time is obtained: Reduced materials handling time and reduced setup time greatly reduce flow time.
- Tool requirements are reduced: Parts produced in a cell are of similar shape, size, and composition. Thus, they often have similar tooling requirements.
- A reduction in space required: Reductions in WIP, finished goods inventories, and lot sizes lead to less space required.
- Throughput times are reduced: In a job shop, parts are transferred between machines in batches. However, in CM each part is transferred immediately to the next machine after it has been processed. Thus, the waiting time is reduced substantially.
- Product quality is improved: Since parts travel from one station to another as single unit, they are completely processed in a small area. The feedback is immediate and the process can be stopped when things go wrong.
- Better overall control of operations: In a job shop, parts may have to travel through the entire shop. Scheduling and material control are complicated. In CM, the manufacturing facility is broken down into manufacturing cells and each part travels with a single cell, resulting in easier scheduling and control.

The basic differences between functional layout and cellular layout is depicted in table.

Table: Differences between functional layout and cellular layout.

| Dimension | Functional | Cellular |
|---|---|---|
| Number of moves between departments | many | few |
| Travel distances | longer | shorter |
| Travel paths | variable | fixed |
| Job waiting times | greater | shorter |
| Throughput time | higher | lower |
| Amount of work in process | higher | lower |
| Supervision difficulty | higher | lower |
| Scheduling complexity | higher | lower |
| Equipment utilization | lower | higher |

## Flexible Manufacturing System

A FMS integrates all major elements of manufacturing into a highly automated system. The flexibility of FMS is such that it can handle a variety of part configurations and produce them in any order. Figure shows flexible manufacturing system. The basic elements of FMS are: a) works station b) automated material handling and automated storage and retrieval systems c) control systems.

Because of major capital investment; efficient machine utilization is essential. Consequently proper scheduling and process planning are crucial, that are complex in nature. Because of the flexibility in FMS, no setup time is wasted in switching between manufacturing operations; the system is capable of different operations in different orders and on different machines.

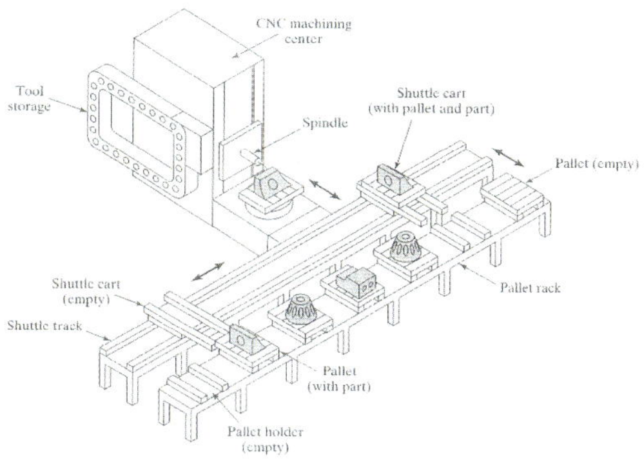


Figure: Flexible manufacturing system

## Advantages:

- Parts can be produced randomly in batch sizes, as small as one, and at lower cost.
- The lead times required for product changes are shorter
- Reduced WIP
- Labour and inventories are reduced

- Production is more reliable, because the system is self correcting and so product quality is uniform.
- Increased machine utilization
- Fewer machines required
- Reduced factory floor space
- Greater responsiveness to change

## Factory System

Modern industry means "Factory system", and can be described as the system of production in which hand tools have been replaced by complicated and costly automatic machines. The works are carried out in large workshops by a large number of workers.

The different requirements of house which could not be fulfilled by one man alone, gave birth to a united work which in turn took the shape of a "Factory". Modern Industries brought the several processes of manufacture under one roof, centralized and increased the use of power, introduced specialized tools and machines and hired workers for fixed wages and hours.

The growth of industry during the past 250 years was specially marked by technological development; the greater increase in inventions of products and processes, the wider use of steam engines as prime-movers, the application of mechanical science to industrial processes, greater specialisation in tools, machines and the labour and rapid expansion of the principle of interchangeable parts.

### Evolution of Factory System

The invention of one machine after another in the 18th and 19th centuries had a very great impact on the economic life of whole world. Actually, the process which first started in England later on spread over the different parts of the world. The system of production which came into being as a result of this "Industrial Revolution" is now known as "Factory System".

### Impact of Factory System on Labour and Management

The modern system of manufacturing has given rise to many problems of various nature, such as:

1. Problem of using complicated machinery.
2. Problem of raising capital for machinery, building and materials.
3. Problem of managing and running the industries.
4. The safety and security of labour class who work in factories and depend on wages for their livelihood has given rise to many social, political and economic problems.
5. The problem of rationalization and modernization.

6. The problem of large requirement of well-trained workers.
7. The problem of competition with foreign market for the purpose of increasing exports.
8. The problem of regional disparity in industries.

## Economic Development of Factory System

From its very beginning, the whole world has been engaged in a continuous social, techno logical and scientific revolution. It has moved from, handicraft to mechanical to automatic mass production, from water power to steam to nuclear power; through mineral, chemical, metallurgical and electronic revolutions.

The factory system may be defined as "an attempt to organise the industries on modern and scientific lines". The development of industries on modern lines first took place in England in 1770.

The economic development of industries can roughly be classified in the following three stages:

### (a) Family or Household System

This is the oldest form. In this each family provides its own requirement (needs). This is the system of making the things in and for a household from home furnished materials. This method today assumes a minor position. The woman who crochets'a table cloth at home from yarn is the example of this system.

### (b) Handicraft System

The family system later on developed into handicraft system. The main feature of this system is specialisation on some particular work such as carpenter, mason, black-smith etc. In this system, one person satisfies a particular need of other and exchanges his labour or product for those of others.

There is a greater differentiation or division of work in this stage than that in the family or household system. The handicraftsman works for the customer. Generally, he has local market in his town and its immediate neighbourhood. A good example of such production is the local tailor who will make you a suit on your order.

### (c) Factory System

This is known as "Modern System" of manufacturing. It satisfies the economic wants of individuals, communities and nations by manufacturing things in workshops which utilize the men, materials, machines, money and methods to meet the market needs.

## Technological Development of Factory System

## Government Policy

After independence Government of India is paying very much attention on industrial and technological development, for which a clear policy was framed in March 1958. Since then, efforts are being made to establish national laboratories and research institutions.

Several such laboratories and research institutions are now working in this field under the control of Council of Scientific and Industrial Research. Several laboratories are running in the different universities.

Atomic Energy Commission is implementing the programme for the development of atomic energy for peaceful purposes. The aim of this programme is to encourage the use of atomic energy in agriculture, biology, industry, medicine, electric generation, mining, irrigation etc.

For the advancement of technical training and research to meet the requirement of technical personnel for the economic development, Engineering Colleges and Polytechnics have been established. In addition, several industrial training centres are training the craftsmen in different trades.

Thus, an atmosphere of science and technology is being created in the country. Now-a-days our country is improving its position considerably in the foreign market.

## Technological Change

Technological development is bringing the change in the world very rapidly. It is playing an important role in improving the living conditions. With the industrial revolution development of means of transport, communication and agricultural techniques have taken place, which resulted certain fundamental changes in social structure.

## Fields of Technical Changes

Following are some of the important fields of technical changes, which have effected the social life considerably:

### (i) Mechanisation of Industries

Mechanisation of industries due to technical development effects the society in different ways, some of them are described hereunder:

(a) Factory System

Mechanisation of industries has given birth to factory system. In this system in place of individual production or job production, integrated plants are installed for mass production. Thousands of people are employed in big industries and high managerial techniques are required to control them. Now-a-days the labour control is becoming a problem with rapid switching over to such industries for high production.

(b) Urbanisation

Big factories are main reason for the growth of big cities. People in-search of work migrates from rural areas to the sites of work and settle there. As a result market centre, schools, colleges, hospitals etc. are established. With the growth of industrial production and social life of the people working in it, new businesses are expected, which result in big business houses, banks, transport companies come to set up and thus form a township.

(c) Rise to New Class

Due to industrialisation new classes came into existence (i) Capitalists who invested money and controlled production, (ii) Workers-who produced for their masters and (iii) then, in between these two there were middle classes and white collared bureaucrats or managers.

Later on conflict started between these classes. Labour class formed the trade unions, while the capitalists formed their federations and thus each wanted to promote their class interests.

(d) Easy Working

In addition to machines for industrial production technology has also brought into use, lots of mechanical appliances through which work at home became easier and less tire-some. These appliances save large amount of time specially for the house wives. Due to these facilities and better standard of living women came out of home to work side by side with men today and started competing with men.

### (ii) Development in the Means of Transport

The means of transport have also developed considerably with the development of technology. Development of cycle, motor cycle, motor car, railway trams, ships and aeroplanes represent the different stages in the growth of transportation.

With the help of these, transportation has become much easier and more quick. They encouraged international trade, which resulted in intermixing of the people belonging to various countries and thereby increasing feeling of sympathy and mutual co-operation.

### (iii) Development in the Means of Communication

Means of communication have been developed through telegraph, telephones, wireless, newspaper, radio and television. These led to mutual exchanges between different cultures.

### (iv) Introduction of New Agricultural Techniques

Quality and quantity of agricultural production improved with the use of new agricultural tools, (like tractors, water pumps, threshers etc.) fertilizers and better seeds. Now agriculture is becoming a commercial occupation and is no more substance activity of the farmer and his family.

The result of these new techniques is to increase an all-round increase in production with fewer people, which resulted the people to shift themselves from farm to the factories or to the trade, in search of other jobs.

## Merits of Factory System

1. Better quality products are available.
2. Wider varieties are manufactured.
3. Less wastage.

4. More efficient employment of land, labour and capital.
5. Higher standard of living.
6. Less unit cost.
7. Increased employment as demand of products increases rapidly.
8. Cultural and social development of workers due to centralization of most of the industries at a place.
9. It reduces training period.
10. Higher rate of production.
11. It yields by-products, which can be used to produce subsidiary products.
12. Goods receive revenue in the form of taxes, duties etc.
13. It helps to a very great extent in defence production.
14. It helps in the advancement of nation as a whole.
15. It helps in the development of international trade.

### Demerits of Factory System

1. It creates monopoly of work.
2. Responsibility of work cannot be fixed on a particular individual easily.
3. It causes over-crowding, unhygienic atmosphere, low wages etc. This all is known as "Evils of factory system".
4. It results in the concentration of wealth in few hands.
5. It causes exploitation of labour and class conflict.
6. It needs large amounts to be spent on competition and advertisement.
7. It may cause over-production.
8. Cottage and small scale industries may suffer.

## Manufacturing Execution System

A manufacturing execution system (MES) is an information system that connects, monitors and controls complex manufacturing systems and data flows on the factory floor. The main goal of an MES is to ensure effective execution of the manufacturing operations and improve production output.

An MES helps achieve that goal by tracking and gathering accurate, real-time data about the complete production lifecycle, beginning with order release until the product delivery stage for finished goods.

The MES collects data about product genealogy, performance, traceability, material management and work in progress (WIP) and other plant activities as they occur. This data, in turn, allows decision-makers to understand the current settings of the factory floor and better optimize the production process.

## Approaching MES

The best approach to MES is to find out the key business goals that need to be achieved and then work backward. This avoids the endless wading through technical specifications that provide a myriad of functionalities that may appear confusing until the context of what is needed is clarified. In reality, many MES functions may be irrelevant for the immediate specific need, and so this helps a great deal with the selection process, though of course considerations need to be remembered for potential future phases.

The selection of MES software and supporting hardware technologies should pave the way towards a single platform that will meet longer-term targeted goals without the need for replacement. The cost of an MES system is not simply the cost of software and hardware but is mainly the investment in the changes that an MES driven operation will make to allow the MES to work effectively. Different MES systems effectively prescribe a "best practice", the fundamentals of which, when compared to both the current and targeted operation, must be considered.

The cost of change and cost of ownership can be more significant than the capital purchase.

## MES and ERP

MES will sit together with other existing and evolve factory control systems. Almost every factory has an enterprise resource planning solution, which in many cases appears to have overlapping functionality with MES. ERP however, typically is orientated towards logical planning for the factory, whereas MES is more concerned with the physical operation. ERP by itself does not have clear knowledge about individual manufacturing processes and capabilities, such that there is an intermediate process required to "translate" the ERP plan into something that can be executed, which without MES this is most often done manually. With MES, information from ERP regarding requirements is used as the basis with which to create specific production operations that MES will then manage. MES, in turn, provides information of greater accuracy about operations back to ERP, which can be used to enhance the abilities to plan next time around.

The relationship between ERP and MES is a key factor for consideration, including where is the point of exchange between the two systems. The simplest way is to define the roles of the two systems according to their strengths, which is most often that the planning is done simply and logically within ERP, while the physical tasks that take place in production are managed by MES in a live and detailed way.

## Starting Out

The starting point when considering MES is the endpoint of production, the completion of products ready for shipping. The critical metric here from the customer perspective is on-time delivery.

For the manufacturing operation, it is the cost of manufacture as well as the capacity of the facility. Capacity is dictated not only by the range of equipment in place but also how it is utilized when faced with the inevitable requirement of changing from one product model to another. The degree of flexibility of both automated and manual production operations dictates the overall effectiveness of the facility when making a high mix of products.

To determine the overall capability to meet delivery goals where many product types are being produced simultaneously is a very difficult calculation. This is the first potential value of MES. Having a computer-based digitalization creates visibility of the status and performance of processes at all times and in all situations, brings opportunities for improvement in the flow of production and even in the assignment of work-orders.

The live tracking of final products and sub-assemblies is a cornerstone of MES. A unique ID, as a label or for example, an etched barcode, can be used to track each production unit on its journey through the factory.

## Simple MES Operation

The unique ID of each production unit is read at each key station, providing several key related values. Firstly, MES processes information about each event, in this case being able to show the location of every production unit. Using the timings of each reading, MES can show visualizations of the flow of production units regarding graphs showing representations of "Key Points of Interest" (KPIs) that help production management and engineering understand the performance of manufacturing. For example, bottleneck processes can be identified where products are queuing before a workstation. Starvation of operations can be found where operators are idle as they wait for the next production unit to arrive. Production units that fail test or inspection processes and need re-routing to repair stations can also be seen, as can the disruption that it causes to the main production flow. Details of efficiency, productivity, Overall Equipment Efficiency (OEE), etc. can all be determined from analysis of the throughput of production units through each production process and travel time between them.

The tracking of production units in this way leads to other benefits. MES can also ensure that when the ID of a product is read, conformance against work-order is assured, such that each production unit can only go through each process in the correct sequence, including consideration of repair loops. This ensures that no production process, or more crucially, a test or inspection operation, is missed. MES systems that are linked with engineering data preparation systems can also ensure for example that the appropriate operation standard and setup documentation for operators will exactly match the product being made, including, for example, any revision variation or material substitutions.

The ability to show documentation electronically eliminates several opportunities for issues associated with paper-based management. Data from automated and semi-automated production processes, such as a test result, can also be collected and attributed against each production unit so that a history of production can be made in the form of production unit traceability. The complexity of this depends on how much data can be obtained from machines and how much time is available for manual operator input. The machine connectivity and user-interfaces are a critical part of the success of MES. The ability to interpret data from many different machines and the ability to gather manual operator input in a consistent and managed way are clear differentiators of MES capability. With manual processes, MES interfaces are personal to the operator, rather than necessarily fixed to an operational location. It is quite usual in manufacturing that there are dedicated physical processes to certain products, for example, functional test, such that not all processes need to be manned simultaneously. The ability of MES to function in a mobile environment rather than being installed at fixed stations is critical in today's dynamic factories.

This simple "first phase" look at the top level of MES provides fundamental visibility of the operation, knowledge about how it is working while highlighting where improvements could be made. There are however many underlying factors that need to be considered to work out how to approach any changes, and what effect those changes may have. Production has many dependencies on other aspects of the total factory operations. These include material preparation and logistics, quality management, engineering data preparation, management of key tools, resources, and of course people. The scope of MES therefore normally goes into deeper aspects of factory management to include support for these dependent processes, such that causes of symptoms seen in the production flow can be addressed, as well as the optimization and improvement of the dependent processes themselves

## Benefits of a Manufacturing Execution System

An MES is often integrated with ERP, supply chain management, product lifecycle management and other key IT systems. A key benefit of using an MES includes:

- Increased customer satisfaction
- Improved regulatory compliance
- Better agility and time to market
- Improved supply chain visibility
- Reduced manufacturing cycle time
- Elimination of paperwork and manual data-entry processes
- Reduced order lead time
- Lower labor costs
- Reduced WIP inventory
- Increased machine utilization

## Computer-integrated Manufacturing

Computer integrated manufacturing (CIM) encompasses the entire range of product development and manufacturing activities with all the functions being carried out with the help of dedicated software packages. The data required for various functions are passed from one application software to another in a seamless manner. For example, the product data is created during design. This data has to be transferred from the modeling software to manufacturing software without any loss of data. CIM uses a common database wherever feasible and communication technologies to integrate design, manufacturing and associated business functions that combine the automated segments of a factory or a manufacturing facility. CIM reduces the human component of manufacturing and thereby relieves the process of its slow, expensive and error-prone component. CIM stands for a holistic and methodological approach to the activities of the manufacturing enterprise in order to achieve vast improvement in its performance.

Manufacturing engineers are required to achieve the following objectives to be competitive in a global context.

- Reduction in inventory
- Lower the cost of the product
- Reduce waste
- Improve quality

Increase flexibility in manufacturing to achieve immediate and rapid response to:

- Product changes
- Production changes
- Process change
- Equipment change
- Change of personnel

CIM is recognized as Islands of Automation. They are:

- CAD/CAM/CAE/GT
- Manufacturing Planning and Control.
- Factory Automation
- General Business Management

CASA/SME's CIM wheel is as shown in figure.

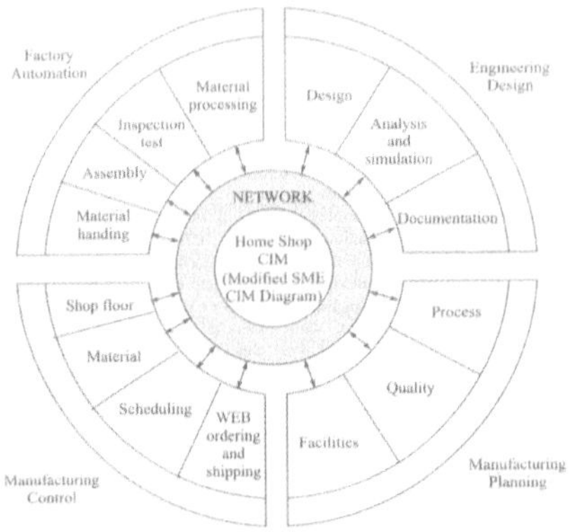


Figure: CASA/SME's CIM Wheel

Computer Integrated Manufacturing (CIM) is considered a natural evolution of the technology of CAD/CAM which by itself evolved by the integration of CAD and CAM. Massachusetts Institute of Technology (MIT, USA) is credited with pioneering the development in both CAD and CAM.

If we review the manufacturing scenario during 80's we will find that the manufacturing is characterized by a few islands of automation. In the case of design, the task is well automated. In the case of manufacture, CNC machines, DNC systems, FMC, FMS etc provide tightly controlled automation systems. Similarly computer control has been implemented in several areas like manufacturing resource planning, accounting, sales, marketing and purchase. Yet the full potential of computerization could not be obtained unless all the segments of manufacturing are integrated, permitting the transfer of data across various functional modules. This realization led to the concept of computer integrated manufacturing. Thus the implementation of CIM required the development of whole lot of computer technologies related to hardware and software.

## CIM Hardware and CIM Software

CIM Hardware comprises the following:

I. Manufacturing equipment such as CNC machines or computerized work centers, robotic work cells, DNC/FMS systems, work handling and tool handling devices, storage devices, sensors, shop floor data collection devices, inspection machines etc.

II. Computers, controllers, CAD/CAM systems, workstations / terminals, data entry terminals, bar code readers, RFID tags, printers, plotters and other peripheral devices, modems, cables, connectors etc.,

CIM software comprises computer programs to carry out the following functions:

- Management Information System
- Job Tracking
- Sales
- Inventory Control
- Marketing
- Shop Floor Data Collection
- Finance
- Order Entry
- Database Management
- Materials Handling
- Modeling and Design
- Device Drivers

- Analysis
- Process Planning
- Simulation
- Manufacturing Facilities Planning
- Communications
- Work Flow Automation

## Elements of CIM System

Nine major elements of a CIM system are in figure they are:

- Marketing
- Product Design
- Planning
- Purchase
- Manufacturing Engineering
- Factory Automation Hardware
- Warehousing
- Logistics and Supply Chain Management
- Finance
- Information Management

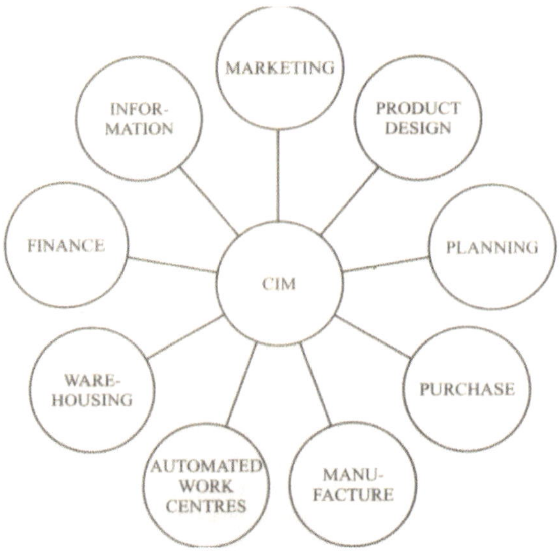


Figure: Major elements of CIM systems

i. Marketing: The need for a product is identified by the marketing division. The specifications of the product, the projection of manufacturing quantities and the strategy for marketing the product are also decided by the marketing department. Marketing also works out the manufacturing costs to assess the economic viability of the product.

ii. Product Design: The design department of the company establishes the initial database for production of a proposed product. In a CIM system this is accomplished through activities such as geometric modeling and computer aided design while considering the product requirements and concepts generated by the creativity of the design engineer. Configuration management is an important activity in many designs. Complex designs are usually carried out by several teams working simultaneously, located often in different parts of the world. The design process is constrained by the costs that will be incurred in actual production and by the capabilities of the available production equipment and processes. The design process creates the database required to manufacture the part.

iii. Planning: The planning department takes the database established by the design department and enriches it with production data and information to produce a plan for the production of the product. Planning involves several subsystems dealing with materials, facility, process, tools, manpower, capacity, scheduling, outsourcing, assembly, inspection, logistics etc. In a CIM system, this planning process should be constrained by the production costs and by the production equipment and process capability, in order to generate an optimized plan.

iv. Purchase: The purchase departments is responsible for placing the purchase orders and follow up, ensure quality in the production process of the vendor, receive the items, arrange for inspection and supply the items to the stores or arrange timely delivery depending on the production schedule for eventual supply to manufacture andassembly.

v. Manufacturing Engineering: Manufacturing Engineering is the activity of carrying out the production of the product, involving further enrichment of the database with performance data and information about the production equipment and processes. In CIM, this requires activities like CNC programming, simulation and computer aided scheduling of the production activity. This should include online dynamic scheduling and control based on the real time performance of the equipment and processes to assure continuous production activity. Often, the need to meet fluctuating market demand requires the manufacturing system flexible and agile.

vi. Factory Automation Hardware: Factory automation equipment further enriches the database with equipment and process data, resident either in the operator or the equipment to carry out the production process. In CIM system this consists of computer controlled process machinery such as CNC machine tools, flexible manufacturing systems (FMS), Computer controlled robots, material handling systems, computer controlled assembly systems, flexibly automated inspection systems and so on.

vii. Warehousing: Warehousing is the function involving storage and retrieval of raw materials, components, finished goods as well as shipment of items. In today's complex outsourcing scenario and the need for just-in-time supply of components and subsystems, logistics and supply chain management assume great importance.

viii. Finance: Finance deals with the resources pertaining to money. Planning of investment, working capital, and cash flow control, realization of receipts, accounting and allocation of funds are the major tasks of the finance departments.

ix. Information Management: Information management is perhaps one of the crucial tasks in CIM. This involves master production scheduling, database management, communication, manufacturing systems integration and management information systems.

## Process

Stage One - Computer aided design. A product is designed totally on computer. When complete it is tested or its functions simulated on screen before even a prototype is made. If a circuit is involved it is designed by using software and tested on screen. Improvements / alterations are made to the design using the same CAD software.

Stage two - prototype manufacture. Prototypes are manufactured on machines such as 3D printers which produce an accurate 3D model. CNC routers and laser cutters may also be used to produce a realistic model. Sometimes working models are manufactured.

Stage three - The computer system controlling the plant works out the most efficient method of manufacture. It calculates costs, production methods, numbers to be manufactured, storage and distribution.

Stage four - The computer system orders the necessary materials to manufacture the product. Keeping costs to a minimum. The 'just in time' philosophy is applied. This means that materials / components are ordered as needed. Very little is stored at the factory. Usually only enough materials are stored to keep the factory going for a small number of days. Materials are automatically reordered when required, to keep the factory working smoothly and continuously.

Stage five - Manufacturing begins with the product being made using CAM (Computer Aided Manufacture). Computers control CNC machines such as laser cutters, CNC routers and CNC lathes.

Stage six - Quality control is applied at every stage. The product is tested using computer control inspections. For instance, the accuracy of manufacture is tested automatically. This ensures that the product is manufactured to the correct sizes.

Stage seven - The product is assembled by robots. This is automated (controlled) by the computer system.

Stage eight - The product is quality checked before being stored for distribution to the customer. All storage is automated. This means that computer controlled vehicles move the finished product from the manufacturing area to storage. The computer systems keep track of every individual product. Products are bar coded which are constantly scanned and recorded by the computer system.

Stage Nine - The product is automatically moved from store to awaiting lorries / trucks for distribution to the customer.

Stage Ten - Financial accounts are updated, bills chased up and paid by the computer system.

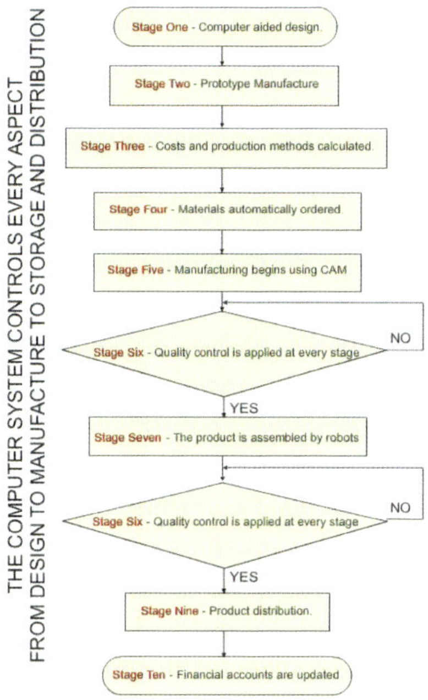


## Flexible Manufacturing System

A Flexible Manufacturing System(FMS) is a manufacturing system in which there is a certain degree of flexibility that allows the system to react in the case of changes, whether predicted or unpredicted. Flexibility is the speed at which a system can react to and accommodate change. To be considered flexible, the flexibility must exist during the entire life cycle of a product, from design to manufacturing to distribution. Flexible Manufacturing System is a computer-controlled system that can produce a variety of parts or products in any order, without the time-consuming task of changing machine setups.

The flexibility being talked about is generally considered to fall into two categories, which both contain numerous subcategories. The first category, Machine Flexibility, covers the system's ability to be changed to produce new product types, and ability to change the order of operations executed on a part. The second category is called Routing Flexibility, which consists of the ability to use multiple machines to perform the same operation on a part, as well as the system's ability to absorb large-scale changes, such as in volume, capacity, or capability.

The main advantage of an Flexible Manufacturing System is its high flexibility in managing manufacturing resources like time and effort in order to manufacture a new product. The best application of an FMS is found in the production of small sets of products like those from a mass production.

FM systems are supposed to provide the manufacturer with efficient flexible machines that increase productivity and produce quality parts. However, FM systems are not the answer to all manufacturers' problems. The level of flexibility is limited to the technological abilities of the FM systems. FM systems are being used all over the manufacturing world and though out industries. A basic knowledge of this kind of technology is very important because FM systems are involved in almost everything that you come in contact with in today's world. From the coffee maker to your remote control FM systems are used all over.

At the turn of the twentieth century, Flexible Manufacturing System did not exist. There was no pressing need for efficiency because the markets were national and there was no foreign competition. Manufacturers could tell the consumers what to buy. During that period, Henry Ford had been quoted as saying "People can order any color of car as long as it is black." All the power remained in the hands of the manufacturer and the consumers hardly had any choices.

However, after the Second World War a new era in manufacturing was to come. The discovery of new materials and production techniques increased quality and productivity. The war led to the emergence of open foreign markets and new competition. The focus of the market shifted from manufacturer to consumer. The first Flexible Manufacturing System was patented in 1965 by Theo Williamson who made numerically controlled equipment. Examples of numerically controlled equipment are like CNC lathes or mills which are varying types of FM systems.

Flexible manufacturing system incorporates the following concepts and skills in an automated production system.

1. Flexible automation
2. Group technology
3. Computer numerical control machine tools
4. Automated material handling between machines

## Types of FMS

Flexible manufacturing systems can be separated into various types subject to their natures:

### Depending upon Kinds of Operation

Flexible manufacturing system can be illustrious subject to the kinds of operation performed:

a. Processing operation. It performs some activities on a given job. Such activities convert the job from one shape to another continuous up to the final product. It enhances significance by altering the geometry, features or appearance of the initial materials.

b. Assembly operation. It comprises an assembly of two or more parts to make a new component which is called an assembly/subassembly. The subassemblies which are joined permanently use processes like welding, brazing, soldering, adhesive bonding, rivets, press fitting.

### Based on Number of Machines

There are typical varieties of FMS based on the number of machines in the system:

a. Single machine cell (SMC). It consists of completely automated machines which are capable of performing unattended operations within a time period lengthier than one complete machine cycle. It is skilful of dispensing various part mix, reacting to fluctuations in manufacture plan, and inviting introduction of a part as a new entry. It is a sequence dependent production system.

b. Flexible manufacturing cell (FMC). It entails two or three dispensing workstations and a material handling system. The material handling system is linked to a load/unload station. It is a simultaneous production system.

c. An Flexible Manufacturing System (FMS). "It has four or more processing work stations (typically CNC machining centers or turning centers) connected mechanically by a common part handling system and automatically by a distributed computer system. It also includes non-processing work stations that support production but do not directly participate in it e.g., part / pallet washing stations, co-ordinate measuring machines. These features significantly differentiate it from Flexible manufacturing cell (FMC).

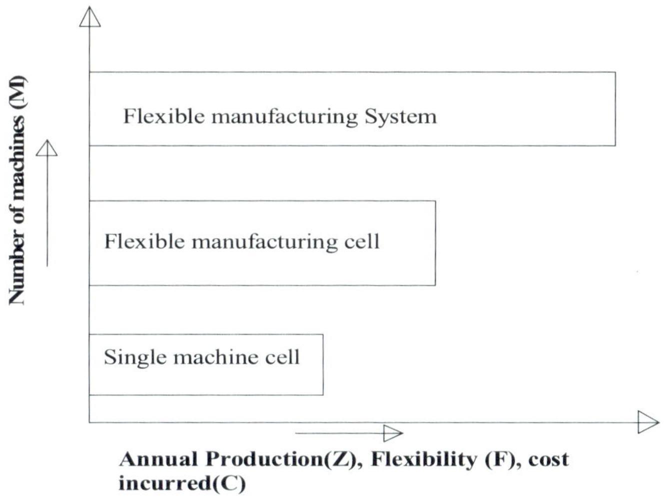


Figure: Comparison for three categories of FMS In this research, authors focused on Flexible manufacturing system

## Based on Level of Flexibility

FMS is further classified based on the level of flexibility related to the manufacturing system. Two categories are depicted here:

a. Dedicated FMS. It is made to produce a certain variety of part styles. The product design is considered fixed. So, the system can be designed with a certain amount of process specialization to make the operation more efficient.

b. Random order FMS. "It is able to handle the substantial variations in part configurations. To accommodate these variations, a random order FMS must be more flexible than the dedicated FMS. A random order FMS is capable of processing parts that have a higher degree of complexity. Thus, to deal with these kinds of complexity, sophisticated computer control system is used for this FMS type.

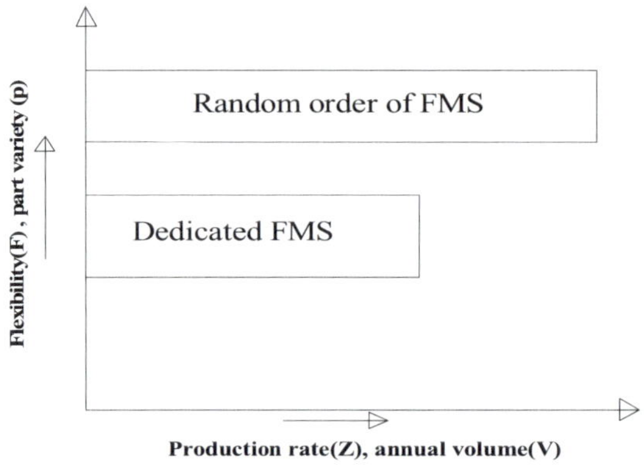


Figure: Differences between dedicated and random-order FMS types

## Elements of Flexible Manufacturing System

A flexible manufacturing system consists of two subsystems:

- Physical subsystem
- Control subsystem

Physical subsystem includes the following elements:

1. Workstations. It consists of NC machines, machine-tools, inspection equipments, loading and unloading operation, and machining area. More recent Flexible manufacturing system, however, include other types of processing equipment also.
2. Storage-retrieval systems. It acts as a buffer during WIP (workin-processes) and holds devices such as carousels used to store parts temporarily between work stations or operations.
3. Material handling systems. It consists of power vehicles, various types of automated material handling equipment such as conveyors, automated guided vehicles, in floor carts and robots are used to transport the work parts and sub-assemblies to the processing or workstation.

## Hardware Components of Flexible Manufacturing System

1. Pallets and fixtures
2. Machining centers
3. Robots
4. Inspection equipment
5. Chip removal system
6. In process storage facility
7. Material handling systems

## Layout Configurations of Flexible Manufacturing System

Flexible manufacturing system brings rewards in actual manufacture of products as the process is designed for several products to be run on different machines within a manufacturing facility which allows for greater growth and stability with more diversity in the output. A Flexible manufacturing system is designed to provide an effective operation sequence to fulfill the production requirements and reasonably allocate the resources. The objectives of the system are to shorten the throughput time and reduce the resource requirements which include avoiding deadlock in material flow, decreasing inprocess inventory, balancing the workload of all machines and make good use of the bottleneck devices.

## Line Layout

An automated guided vehicle is most efficient when the movement is in straight-lines along the AGV path in a single-row machine layout. Machines are arranged only on one side of AGV path, and in double row machine layout, machines are arranged on both sides. A possible arrangement of this layout is shown in figure.

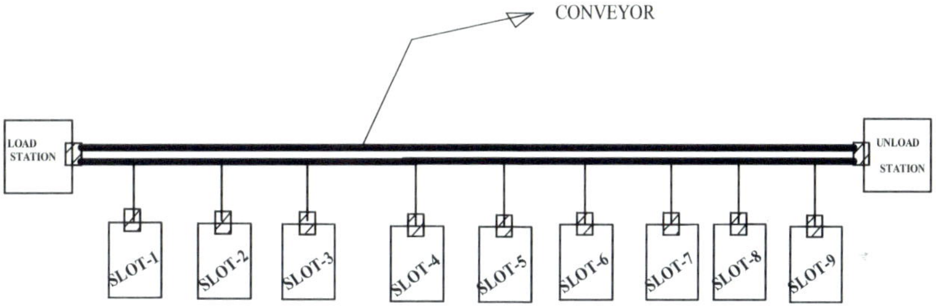


Figure: Line layout

## Loop Layout

The loop layout uses conveyor systems that allow unidirectional flow of parts around the loop. A secondary material handling system is provided at a workstation which permits the flow of parts without any obstruction. A possible arrangement of this layout is shown in figure.

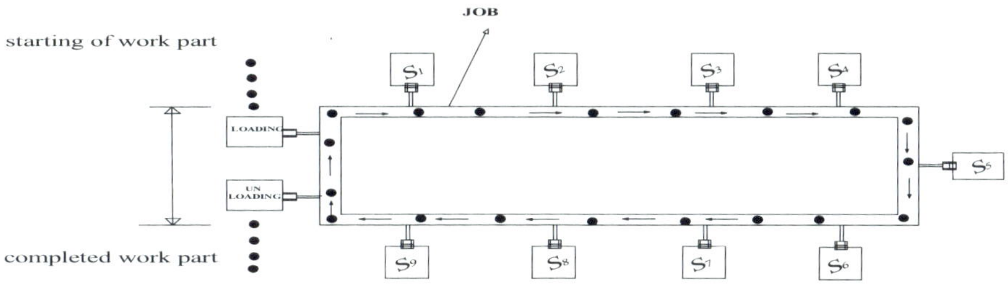


Figure: Loop layout

## Ladder Type Layout

Ladder type layout consists of rungs on which workstations are located. This reduces the average travel distance thereby reducing the transfer time between workstations.

A possible arrangement of this layout is shown in figure.

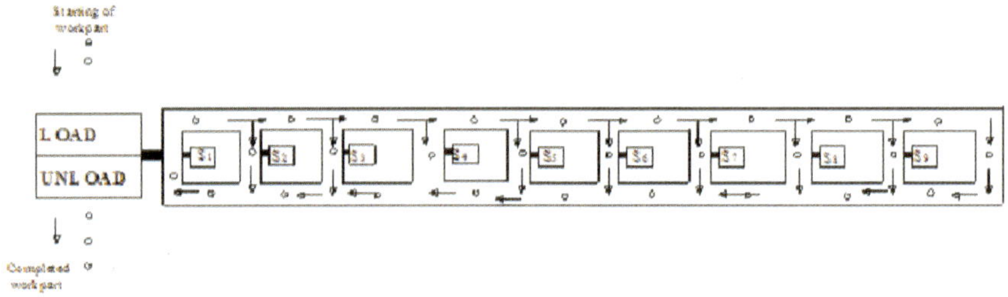


Figure: Ladder layout

## Carousel Layout

In the Carousel layout configuration, parts flow in one direction around the loop. The load, unload stations are placed at one end of loop, A possible arrangement of this layout is shown in figure.

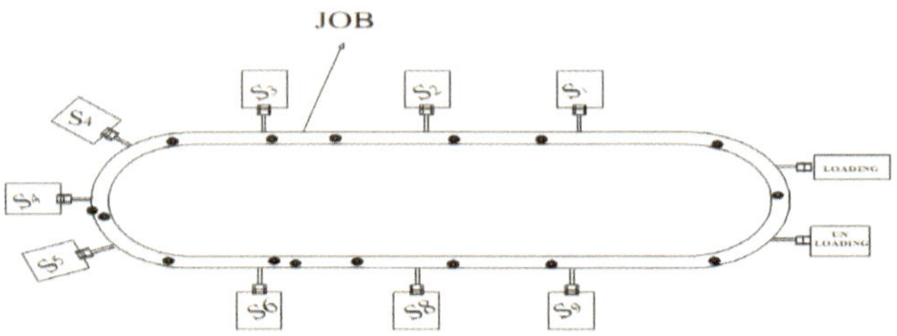


Figure: Carousel layout

## Robot Centered Cell

If a handling robot is used in a Flexible manufacturing system cell, the machines are laid out in a circle, such a layout is called circular layout. A possible arrangement of this layout is shown in figure.

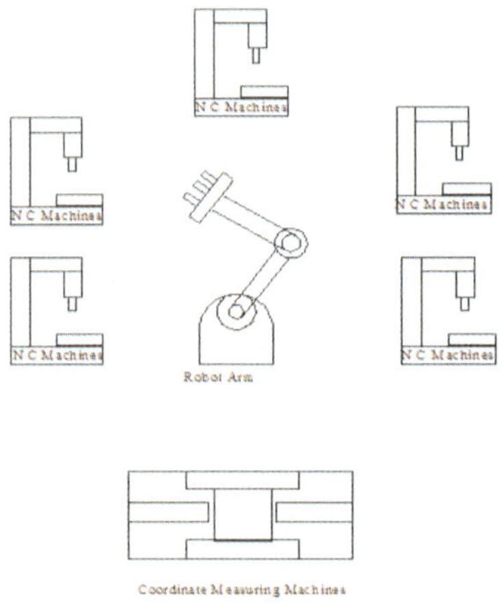


Figure: Robot centered cell

## Open Field Layout

The open field layout is also an adoption of the loop configuration. The open field layout consists of loops and ladders organized to achieve the desired processing requirements. This is used for the processing of a large family of parts. The number of different machines may be limited, and the parts are routed to different workstations depending on availability of machines. A possible arrangement of this layout is shown in figure.

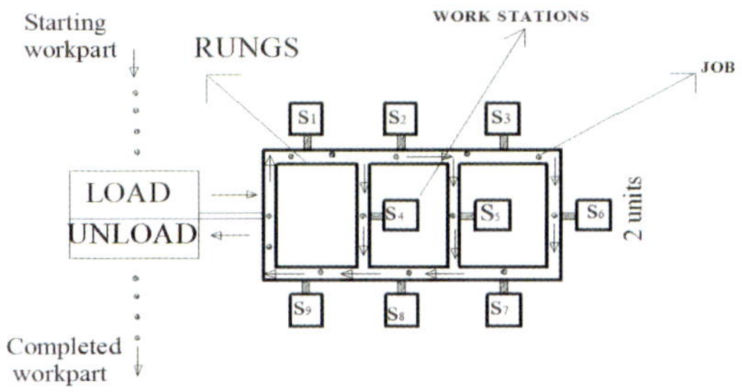


Figure: Open field layout

## Scheduling of Flexible Manufacturing System

## Introduction to FMS Scheduling

Scheduling is a method of totaling start and finish time data to the job sequencing dictated in the sequencing process. Sequencing process in turn, is defined as getting the order in which jobs are processed on a machine. The sequence thus obtained determines the schedule, since authors assume each job is started on the machine as soon as the job has finished all predecessor operations and the machine has completed all earlier jobs in the sequences. This is referred to as a semi-active schedule and acts as an optimal policy for minimizing the completion time, flow time, lateness, tardiness, and other measures of performance. Scheduling problems are often denoted by N/ M/ F/ P, where N is the number of jobs to be scheduled; M is the number of machines; F refers to the job flow pattern and P is performance measures that are to be appropriately minimized or maximized. The solution of scheduling problems is generally presented in the form of gantt-chart which is plotted between different work centers and total processing time on that work center.

Flexible material-handling systems for manufacturing have the capability of moving articles or carrying materials between process stations in different sequences. The traditional method for controlling the routing of carriers is to determine, in advance, all of the useful paths within the system, and store the information in a central computer until needed.

Flexible manufacturing system scheduling could be considered as a static scheduling problem where a fixed set of orders are to be scheduled either using optimization or priority scheduling heuristics. Alternatively this could also be viewed as a dynamic scheduling problem where orders arrive periodically for scheduling as daily orders are released from a material requirement planning system or as individual customer's orders. For these types of scheduling, performance criteria minimizing the make span or the mean order lateness and tardiness are usually considered.

Under both types of scheduling problems, three different scheduling and control decisions are taken while managing day-to-day operations.

i. When to load a part into FMS?

ii. Which type of part to be loaded into FMS?

iii. Dispatching of part.

For the first decision, factors such as system congestion and part fixture availability are considered. The second decision concerns choosing the type of part to enter the FMS at the loading station when it is time for a new part to enter the system, considering such factors as part characteristics and machine workload conditions. The dispatching decision is depend on routing parts through the FMS at the time of actual production, and also sequencing parts at the individual machines in an FMS.

## Functions of Scheduling

To design an effective production scheduling system, the following functions must be carried out in most systematic and efficient manner.

1. Assigning different jobs to different facilities after considering the feasibility of allocation (loading).
2. Establishing a set of rules of priorities to sequence the activities on the facilities.(sequencing)
3. Dispatching job orders as per the schedule to initiate loading of jobs to facilities.

Reviewing the status of jobs i.e., monitoring the achievement and the scheduling or changing the priorities of the jobs to rectify the deviation.

## Objectives of Scheduling

The objectives of FMS scheduling:

1. Achieving high efficiency of the operations by optimally utilizing machines and equipment.
2. Maintaining low inventories in process and raw materials.
3. Maintaining short flow time of products.

The objectives often become conflicting to each other and as such the scheduling process involves a trade–off conflicting objectives such that proper balance is obtained.

## Elements of Scheduling

Scheduling determines the timings and order of the operations to optimize the use of resources to meet production requirements. The following are the elements of scheduling.

a. Job arrival patterns

b. Number and variety of machines /operations

c. Ratio of workers to machines/operations

d. Flow pattern of jobs

e. Priority rules for allocating work

## Principles of Scheduling

In view of the direct equivalence between the workflow and cash flow, scheduling should be done with utmost care. The following are the major principles of scheduling.

1. Effectiveness should be measured by speed of workflow
2. Scheduling jobs as a string with process step back
3. A job started should not be interrupted
4. Speed of flow is most efficiently achieved by focusing on bottlenecks
5. Reschedule every day vi. Obtain feedback on each job at each work center every day
6. Match the work center input information to that the worker can actually do
7. When seeking improvement in output, look for incompatibility between engineering design and process execution
8. Always work to attain certainty of standards and routings

## Scheduling Priority Rules

A sample set of priority rules are presented below

In this work, there are six kinds of rules for operation selecting from the given allowed set.

(1) Short Processing Time (SPT),

- Process the operation with minimum processing time.

(2) Most Work Remaining (MWR),

- Process the operation among the tasks with the maximum total processing time.

(3) Random Selecting Rule,

- Select an operation from the giving allowed set with a random way.

(4) Long Processing Time (LPT),

- Process the operation with maximal processing time.

(5) Most Operation Remaining (MOR),

- Process the operation among the tasks with most remaining operation.

(6) Earliest due date Rule (EDD) rule,

- The earliest due date rule minimises the maximum job lateness, or job tardiness, but unfortunately it tends to need more tasks and increased the mean tardiness

The author made an attempt on a few priority rules such as Most Work Remaining (MWR), Short Processing Time (SPT), Random Selecting Rule in this task of scheduling for optimum arrangement of machines in FMS layout.

### Benefits of FMS

Flexible manufacturing system benefits are listed as follows:

- Reduction of inventories throughout the complete chain of manufacturing including work-in-progress
- Reduction of lead time by 40%
- Improved machine utilization by 30%
- Reduction of labour times by 30%
- Reduction of direct and indirect labour costs
- Increased management control over the entire manufacturing process
- Substantially reduced scrap levels.
- The ability to adapt quickly to new work pieces.

## Reconfigurable Manufacturing System

A reconfigurable manufacturing system is a system used by manufacturers that emphasizes the importance of being able to change and evolve rapidly in order to adjust its productivity capacity and functionality. When a sudden change occurs in the market, the reconfigurable manufacturing system changes in response, allowing the company to produce products or goods in an efficient manner.

The reconfigurable manufacturing systems have two important components:

1. Reconfigurable machine tool,
2. Reconfigurable controller.

### Reconfigurable Machine Tool

The uniqueness of reconfigurable manufacturing system is that the structure of the system as well as of its machines and control can be rapidly changed in response to market changes. A major component of RMS is the reconfigurable machine tools (RMT). By contrast, to conventional CNC those are general purpose machines, RMT are designed for specific customized range of operation requirement and may be cost effectively converted when the requirements change. Reconfigurable machine tools that are designed to produce specific set of features for specific range of cycle time. Some operation requirement will be constant over the life time of the machining system. The primary aim of the reconfigurable machine tool is to cope with various changes in the products or parts to be manufactured. The following possible changes must be taken in to consideration.

- Work piece size
- Part geometry and complexity
- Production volume and production rates
- Required processes
- Accuracy requirement in terms of geometrical accuracy, surface quality
- Material property such as kind of material, hardness etc.

Reconfiguration requirement introduce several new challenges for RMT controllers. The first challenge is the reconfigurable of the controller architecture that is required when the physical machine tools are reconfigured or new technology integrated. Another challenge is the control of RMT with multiple tools working independently and RMT with axes in non orthogonal configuration. The other important challenge is the integration of heterogeneous software and hardware components. The design of RMT requires broad knowledge of machine design, machine tool design, kinematics modeling and dynamics analysis. There is no comprehensive theory or design methodology that is directly applicable to the RMT design. The concept of reconfiguration has been used in related fields including fixture design, assembly system design and reconfigurable robots. Some typical motion types are defined as follows.

## RMT Tool Machining Motion

RMT (tool machining motion) collects all reconfigurable machine tool motions, which are involved in the same machining feature (hole, slot, etc.). All of tool point motions have the same directions (x-axis, y-axis, etc.) and perform the same machining task.

## RMT Tool Machining Motion Family

An RMT TMMF is the collection of a series of machining motions among different work pieces, which can share the same motion types. According to reconfigurable machines, tools have four important modules to develop reconfigurable machine tool control system as follows.

1. Automatic part transfer system control module
2. Automatic part clamping rotating system control module
3. Automatic part lifting system control module
4. Automatic tool changing system control module.

## Reconfigurable Controller

To control a particular machine, any machinespecific functions or classes currently must be designed and built into a Reconfigurable Controller. The Reconfigurable controller becomes unchangeable at run-time for controlling different machines. For instance, reconfigurable controller can control a three-axis tabletop mill and a five-axis mill. To allow reconfiguration of the motion planner and servo controller necessary for controlling different mechanisms, some additions and modifications must be made to reconfigurable controller. The configuration system is directly interfaced so that it can receive configuration commands from this interface. Based on these different configuration instructions, the configuration system will do one of two operations. It will either set up a correct machine operational parameter, such as machine joint limits. After the configuration system finishes all of these configuration processes, the reconfigurable controller is dynamically reconfigured for a particular mechanism. CAD/CAM applications can then pass the manufacturing process instructions to the reconfigurable controller for direct machining.

## Capabilities of Reconfigurable Manufacturing System

The reconfigurable capabilities could be divided into four different main dimensions.

- Product related change capabilities
- Change competency within operations
- Cooperation internally and externally
- People, knowledge, and creativity
- Product Related Change Capabilities

Design for manufacturing is the most important area that plays a vital role to improve the product related capabilities if the products are designed for assembly this will make the system more productive. The products could also be designed for reconfigurablity in the sense of using similar material, redundancy, and efficient change over. The information technology is also the important area, which helps to improve the product related change capabilities. Information technologies have in some examples in industry been used to configure products providing customization in software instead of hardware. Software opens up for both flexibility and reconfigurablity if applied in this way.

## Change Competencies within Operations

In the decision process the aspects of generating alternatives and evaluating are vital for successful decisions. Also, a condition for any decision process is the ability to generate and collect information. The gathering of information, generation of alternatives, and evaluation depends on which problem is being investigated. Thus, creativity and innovations are central to formulating alternatives and solutions. Also a structured approach is vital in the decision making process. This means that flexibility and reconfigurablity in conjunction to production system changes are very closely related to human beings within the production system and cannot be achieved by technology alone.

## Cooperation Internally and Externally

Good product design and methods as design for assembly require good cooperation between design and manufacturing. So that several methodologies of concurrent engineering have been developed, this can be used to improve cooperation. As the production system spans over several actors it becomes more and more important to cooperate with other companies.

Thus, the methodologies of concurrent engineering must also be expanded to include different companies.

## People, Knowledge and Creativity

The last defined dimension of reconfigurable capabilities is people, knowledge, and creativity. There exist numerous ways to develop these capabilities, mainly related to management issues. Motivation can be developed by good management and is in itself a large area for research.

## Characteristics of Reconfigurable Manufacturing Systems

Based on literature review, various types of characteristics of Reconfigurable Manufacturing System as follows:

## Dynamic Behavior

The dynamic behavior may be described from two senses, namely the reactive activities and the proactive activities. Adaptability of manufacturing systems mainly refers activities reactively adopted by a manufacturing system to adapt to environmental changes. High adaptability of a reconfigurable manufacturing system permits quick and cost-effective response to changes. The reconfigurable manufacturing system will not only adapt to changes, it also has to be robust against some external changes, so they can keep a needed degree of stability for periods.

## Integrated Information

Manufacturing systems are able to select the necessary data from the data-overloaded environment and to filter the right information. This will be a major core competence of a system to guarantee its own survives in a fast changing and fast moving environment. But beside the management of a huge amount of data, there is also the necessity for systems to ensure that the right persons get the needed information in a proper quality and also in time. Information integration meets requirements of communication and information sharing among autonomous units. Therefore, information integration is strongly related to intelligence and decision-making capabilities of reconfigurable organization units.

## Oriented Innovation

Innovations will be one of the best characteristics supporting manufacturing reconfigurablity concerning products, processes, organizations and markets. There may be different kinds of innovations, such as technological or social innovations. Social innovations influence the behavior of the people within manufacturing systems and create an environment where people's imagination and creativity are encouraged. On the other hand, technology innovations are the activities that create new market demands.

## Environment Consciousness

Reconfigurable manufacturing systems have to be environment-conscious. It takes the ecological concerns in the whole being of manufacturing systems and products into account. Reconfigurability of manufacturing systems and products will reduce and avoid negative impacts on the environment to ensure a sustainable development.

## Competition and Cooperation

Competition between manufacturing systems can be identified as another enabler to force the reconfigurablity of systems. It is one factor that leads to internal adaptation to be more competitive. Cooperation can be seen as a characteristic that causes internal changes but mainly influences the market and the behavior of the competitors. One can identify mainly vertical or horizontal cooperation between non-competing enterprises. But environmental constraints also force the cooperation between direct competitors.

## Challenges of Reconfigurable Manufacturing System

The concept of reconfiguration has sparked interest in the academic and industrial communities. It has encouraged active research into supportive areas that are proving very beneficial to existing manufacturing systems, e.g., in the areas of process and production planning, fixturing, modular interfaces, and the like. Technology is available today to achieve a useful and affordable, physical and logical reconfiguration within manufacturing systems until new technologies are developed. The several challenges of reconfigurable manufacturing system as follows:

1. Measures for changeability, flexibility, adaptability, responsiveness, reconfigurability and their relationships.
2. The hardware and software enabling technologies,
3. Reconfigurable logical support systems, such as logistics, production planning and control, process planning, tooling, and fixtures,
4. Balance of hard and soft capacity and functionality scalability options,
5. Design of machines, systems, and controls for flexibility, changeability, and reconfiguration and integration with current systems and software,
6. Models to determine adequate levels of changeability, flexibility, and reconfigurability required for different applications,
7. Appropriate capacity scalability policies
8. Appropriate frequency of change or reconfiguration,
9. Smooth and optimal systems transition and changeover,
10. Changeability and reconfiguration dependent quality factors, including human- related issues,
11. Complexity measurement, reduction, and management techniques,
12. The use of group technology to capitalize on commonality and standardization of parts, operation sequences, product structure, platforms, engineering.

# Lean Manufacturing

Lean manufacturing is a methodology that focuses on minimizing waste within manufacturing systems while simultaneously maximizing productivity.

Also known as lean production, or just lean, the integrated sociotechnical approach is based on the Toyota Production System and is still used by that company, as well as myriad others, including Caterpillar Inc. and Nike

Lean manufacturing is based on a number of specific principles, such as Kaizen, or continuous improvement.

Lean manufacturing was introduced to the Western world via the 1990 publication of The Machine That Changed the World, which was based on a five-year, $5 million MIT study of the future of the automobile that detailed Toyota's lean production system. Since that time, the lean principles have profoundly influenced manufacturing concepts throughout the world, as well as industries outside of manufacturing, including healthcare, software development and service industries.

The benefits of lean include reduced lead times, reduced operating costs and improved product quality, to name just a few.

## Five Principles of Lean Manufacturing

A widely referenced book, which was published in 1996, laid out five principles of lean, which many in the field reference as core principles. They are value, the value stream, flow, pull and perfection. These are now used as the basis for lean implementation.

1. Identify value from the customer's perspective. Value is created by the producer, but it is defined by the customer. In other words, companies need to understand the value the customer places on their products and services, which, in turn, can help them determine how much money the customer is willing to pay. The company must strive to eliminate waste and cost from its business processes so that the customer's optimal price can be achieved at the highest profit to the company.

2. Map the value stream. This principle involves recording and analyzing the flow of information or materials required to produce a specific product or service with the intent of identifying waste and methods of improvement. The value stream encompasses the product's entire lifecycle, from raw materials through to disposal. Companies must examine each stage of the cycle for waste - or muda in Japanese. Anything that does not add value must be eliminated. Lean thinking recommends supply chain alignment as part of this effort.

3. Create flow. Eliminate functional barriers and identify ways to improve lead time to ensure the processes are smooth from the time an order is received through to delivery. Flow is critical to the elimination of waste. Lean manufacturing relies on preventing interruptions in the production process and enabling a harmonized and integrated set of processes in which activities move in a constant stream.

4. Establish a pull system. This means you only start new work when there is demand for it. Lean manufacturing uses a pull system instead of a push system. With a push system, used by manufacturing

resource planning (MRP) systems, inventory needs are determined in advance and the product is manufactured to meet that forecast. However, forecasts are typically inaccurate, which can result in swings between too much inventory and not enough, as well as subsequent disrupted schedules and poor customer service. In contrast to MRP, lean manufacturing is based on a pull system in which nothing is bought or made until there is demand. Pull relies on flexibility and communication.

5. Pursue perfection with continual process improvement, or kaizen. Lean manufacturing rests on the concept of continually striving for perfection, which entails targeting the root causes of quality issues and ferreting out and eliminating waste across the value stream.

## Eight Wastes of Lean Production

The Toyota Production System laid out seven wastes, or processes and resources, that don't add value for the customer. These seven wastes are:

- unnecessary transportation;
- excess inventory;
- unnecessary motion of people, equipment or machinery;
- waiting, whether it is people waiting or idle equipment;
- over-production of a product;
- over-processing or putting more time into a product than a customer needs, such as designs that require high-tech machinery for unnecessary features; and
- defects, which require effort and cost for corrections.

Although not originally included in the Toyota Production system, many lean practitioners point to an eighth waste:

- Waste of unused talent and ingenuity.

## Seven Lean Manufacturing Tools and Concepts

Lean manufacturing requires a relentless pursuit of reducing waste. Waste is anything that customers do not believe adds value and for which they are not willing to pay. This requires continuous improvement, which lies at the heart of lean manufacturing.

Other important concepts and processes lean relies on include:

- Heijunka: Production leveling or smoothing that seeks to produce a continuous flow of production, releasing work to the plant at the required rate and avoiding interruptions.
- Kanban: A signal -- either physical, such as tag or empty bin, or electronically sent through a system -- used to streamline processes and create just-in-time delivery.
- Jidoka: A method of providing machines and humans with the ability to detect an abnormality and stop work until it can be corrected.

- Andon: A visual aid, such as a flashing light, that alerts workers to a problem.
- Poka-yoke: A mechanism that safeguards against human error, such as an indicator light that turns on if a necessary step was missed, a sign given when a bolt was tightened the correct number of times or a system that blocks a next step until all the previous steps are completed.
- 5S: A set of practices for organizing workspaces to create efficient, effective and safe areas for workers and which prevent wasted effort and time. 5S emphasizes organization and cleanliness.
- Cycle time: How long it takes to produce a part or complete a process.

## Lean vs. Six Sigma

Both lean and Six Sigma seek to eliminate waste. However, the two use different approaches because they see the root cause of waste differently.

In the simplest terms, whereas lean holds that waste is caused by additional steps, processes and features that a customer doesn't believe adds value and won't pay for, Six Sigma holds that waste results from process variation. Still, the two approaches are complementary and have been combined as Lean Six Sigma.

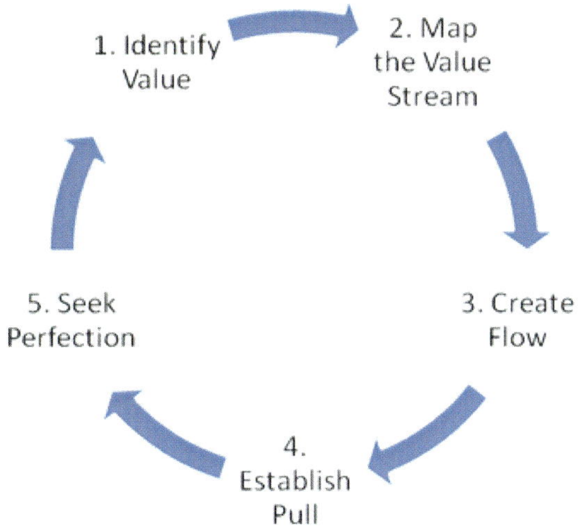


## Agile Manufacturing

Agile manufacturing is a vision of manufacturing that is a development from the concept of Lean production. In lean production, the emphasis is on the elimination of sources of waste. Lean implies high productivity and quality, but it does not necessarily imply being responsive to unique customers' demands. Agile stresses the importance of being highly responsive to meet the `total needs' of the exact customers.

The agile manufacturing system should be able to produce a variety of components in a short time period. To achieve Agile Manufacturing status, companies primarily need implement agile design processes. Agile product development system is capable of addressing frequent iterations of multiple design options early in the process, based on continuous testing and highly sophisticated customer driven design changes. Agile manufacturing suggests certain future directions for the further evolution of the lean enterprise system to help enterprises develop capabilities to thrive in fast-changing and uncertain environmental conditions.

The agile manufacturing enterprise can be defined along four dimensions:

1. Value-based pricing strategies that enrich customers;
2. Co-operation that enhances competitiveness;
3. Organizational mastery of change and uncertainty; and
4. Investments that leverage the impact of people and information.

Agility has four underlying principles:

1. Delivering value to the customers
2. Being ready for change
3. Valuing human knowledge and skills
4. Forming virtual partnerships.

## Key Factors

### Customer Focused Product Design

Consumers expect more personalized attention than ever before. This extends not only to rapid service but also to the products they order. An agile organization will keep on top of market demand and, as well, design the production process so that desired product variations are accounted for in the production planning.

### Fully Connected IT

Quick response to market demand is difficult without an integrated technology environment. Information flow needs to be accurate and in real time. This enables all areas of an agile manufacturing organization to fulfill orders correctly and efficiently. Everyone involved, from sales teams and customer service agents to line workers, needs to be on top of reliable production and market information. This information, of course, must be supported with the ability to act upon it quickly.

### Supply Chain Cooperation

Agile methodologies involve both internal teams and external alliances. Suppliers can only respond quickly when armed with adequate production flow information, and shippers need to know what to expect for rapid delivery to end-users. Strategically, sometimes it is helpful when new alliances include a partner that is involved with some part of the product build-out or design. In sum, your supply chain must be aligned with your customer demand.

### Employee Skill-sets

Employee training for agile involves learning basic information on new production practices in addition to education in a new, agile, customer-driven format. Expecting employees to understand the rapid changes and adaptations they will be involved in requires intensive training and an acknowledgement that a cultural change is occurring.

### Full Company Involvement

An agile transition will not work if it is relegated to the team level. Company structure needs to be part of the transformation to becoming agile. The changeover usually works better if an operation is introducing a new plant or expanding an older one. That way, the focus can be on bringing the entire new area up to agile standards. Requesting workers handle operations as they did previously while expecting them to concentrate on setting up an agile manufacturing module is doomed to fail.

Moving an organization from a lean manufacturing mode to an agile manufacturing company with swift response to customers, offering them what they want when they want it, requires planning. It will be a path involving everyone in the company and it will change the old order dramatically, but it is rewarding, supports growth, and provides a huge competitive advantage to the brave managers that decide to walk down that path.

## Biomanufacturing

Biomanufacturing is a new type of production that uses biological systems to construct commercially-relevant biomaterials to add to medicine, industrial applications and the food and beverage industry. Biomanufactured products are found in natural sources like cultures of microbes, blood or plant and animal cells that have been artificially grown in specialized equipment. The cells used in the production may have occurred naturally or through advanced genetic engineering techniques.

### Biomanufactured Products

Most consumers are unaware that there are thousands of biomanufactured products available on the market today. They are classified by medicine, food and beverage and industrial applications. In the medicine field, amino acids, vaccines, cytokines, fusion proteins, growth factors, biopharmaceuticals and monoclonal antibodies all utilize biomanufactured products. Again, amino acids are biomanufactured for the food and beverage industry to improve the health of the food or drink. Other biomanufactured products found in this category are enzymes and protein supplements.

Finally, industrial applications that require cells or enzymes that are biomanufactured include biocementation, detergents, plastics and bioremediation. Biocementation is microbiologically-induced calcium carbonate precipitation which is commonly used in the treatment of concrete to prolong service life, as a result of the added calcium, as well as in bricks and fillers for plastics and ink. Bioremediation is a technique used in waste management that involves organisms neutralizing pollutants and other contaminants. The organisms break down hazardous substances into much less toxic (or non-toxic) materials.

Methods of Biomanufacturing

While there are many methods of biomanufacturing useful cells, the following are the most popular:

### Blood Plasma Fractionation

Blood plasma fractionation is the process of separating components of blood plasma which is the liquid portion of the blood. It contains numerous proteins such as immunoglobulins and albumin as well as clotting proteins like fibrinogen. These are most commonly used in medicines to improve the immune system.

### Cell Culture

Cell culture is the process by which cells are grown in an artificial environment under controlled conditions. Animal and plant cells are replicated and grown using this method in a culture dish. This is beneficial for medicines, supplements and enzymes.

### Column Chromatography

Column chromatograph is the method used to purify chemical compounds from mixtures. This process is typically used for pharmaceuticals and medicines and offers the benefits of a relatively low cost and easily disposable.

### Fermentation

Fermentation is the process known to produce alcohol. It has functional purposes. Fermentation is a metabolic process converting sugar into gases, acids and alcohol. This is most important for the food and beverage industry as well as with industrial applications.

### Homogenization

Homogenization is the process by which a biological sample is to a point in which all fractions are equal in composition. This is useful in medicine and industrial applications to create a balance amongst the cell constructions.

## Ultrafiltration

The ultrafiltration process involves the separation of solids and solutes with high molecular weights which need to be retained from the process. Low molecular weights pass through the semipermeable membrane. The process is used in drinking water, waste water treatment and other requirements for recycling processes.

## Types of Equipment

The type of equipment and manufacturing facility requirements are dictated by the products being produced. The majority of process equipment is constructed from plastic or stainless steel because their chemical constructions do not interfere with process. Also, stainless steel is easy to clean and can be reused for other functions. Unfortunately some plastic equipment must be disposed of following a single use.

Those products manufactured for food and beverage or medical usage must be produced in a facility that has been designed within the confines of Good Manufacturing Practices (GMP) regulations. In many instances, cleanrooms are required to control the particulates and microorganisms. Also, aseptic processing equipment and sterilization are required for any molecules that will be part of an injectable product.

## References


- Factory-system-meaning-evolution-and-merits-7084: businessmanagementideas.com, Retrieved 26 July 2018
- Manufacturing-execution-system-MES: searcherp.techtarget.com, Retrieved 11 May 2018
- What-is-mes-manufacturing-execution-systems: newcastlesys.com, Retrieved 31 March 2018
- What-is-a-reconfigurable-manufacturing-system: monroeengineering.com, Retrieved 11 June 2018
- What-is-agile-manufacturing: planettogether.com, Retrieved 11 May 2018
- Biomanufacturing: moneyinc.com, Retrieved 28 March 2018

# 2

# Manufacturing Process Management

Manufacturing process management is a set of techniques and methods that are used to guide the manufacturing of products. This chapter has been carefully written to provide an easy understanding of the varied facets of manufacturing process management, such as near net shape, metal casting, subtractive manufacturing, molding, metal forming, etc.

Manufacturing process is basically a complex activity, concerned with people who've a broad number of disciplines and expertise and a wide range of machinery, tools, and equipment with numerous levels of automation, such as computers, robots, and other equipment. Manufacturing pursuits must be receptive to several needs and developments.

Beside above, all the future technicians must understand the basic needs of workshop routines in terms of man, equipment, material, methods, revenue and other infrastructure conveniences needed to be placed properly for maximum shop or plant layouts and other support solutions effectively regulated or positioned in the field or industry within a properly planned manufacturing firm.

The complete knowledge of fundamental workshop technology and manufacturing processes is highly troublesome for anybody to claim competence over it. It deals with numerous aspects of workshops procedures also for providing the basic working awareness of the various engineering materials, tools, accessories, manufacturing processes, basic concepts of machine instruments, production criteria's, traits and uses of numerous testing instruments and calibrating or inspecting units for checking materials or products designed in various production shops in a commercial environment. It also explains and illustrates the use of several hand tools (calibrating, marking, forming and supporting gear etc.), tools, machinery and diverse methods of production that facilitate forming or shaping the existing raw materials into appropriate usable forms. Below are some of the manufacturing processes that are worth reading.

## Classification of Manufacturing Processes

Manufacturing processes can be classified as processing operation and assembly operation.

In processing operation, the work material is transformed from one state to other advanced state. Through this operation value is added to the work material by changing the geometry; shape properties, appearance etc. of the starting work material. Usually processing operations are performed on individual component. But in some cases like aerospace industry, the processing operations are performed on assembled items also.

In assembly operation, two or more components are joined to create a new entity. The new entity is called assembly, subassembly based on its state in the product. If the entity is an intermediate state of the product, it is called subassembly. Some other terms are also referred based on the joining process. The assembly created by welding operation is called weld met.

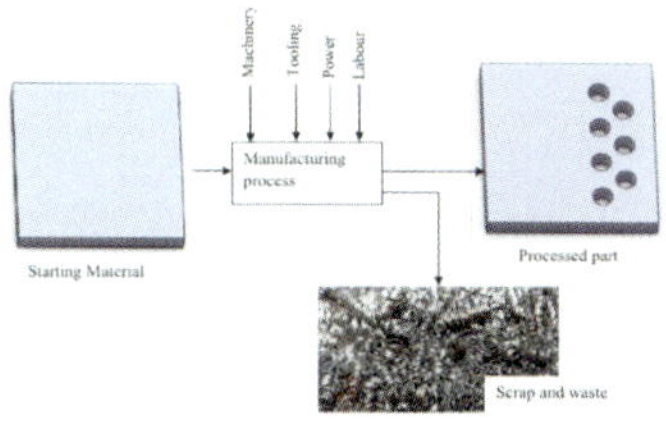


Figure: Definition of manufacturing in terms of technology.

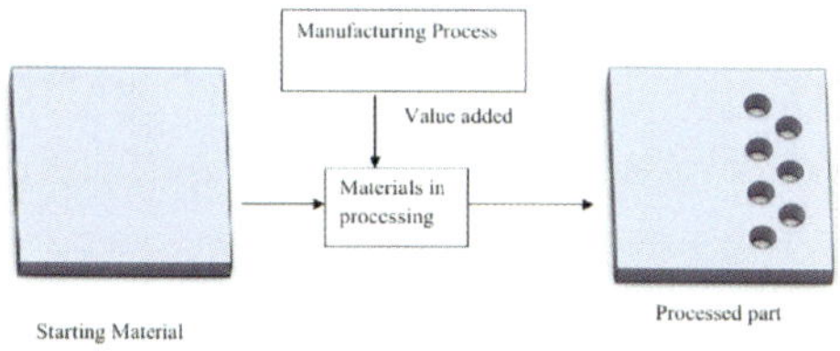


Figure : Definition of manufacturing in terms of economic value.

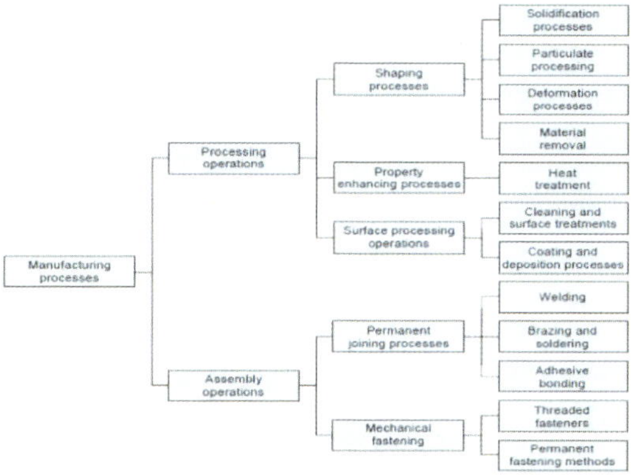


Figure.: Classification of manufacturing process

## Job and Station

In classical manufacturing, a job is defined as the total work or duties a worker performs. A station is a location or area where a production worker performs tasks and jobs. A job at an assembly station may consist of the following tasks:

1) Attach carburetor
2) Connect gas line
3) Connect vacuum line
4) Connect accelerator rod.

Jobs and stations are now applied to unmanned machines also. A simple machine may have only station. A complex machine may have multiple stations.

## Operation

An operation is a distinct action performed to produce a desired result. Operations can be

1) Materials handling and transporting
2) Processing
3) Packaging
4) Inspecting and testing
5) Storing

## Treatments

Treatments operate continually on the work piece. These modify the product in process without tool contact. Some examples include heat treating, curing, galvanizing, plating, finishing, (chemical) cleaning and painting. These are often done in large tanks or furnaces or rooms, away from workers as they can be harmful.

## Basic Manufacturing Processes

It is described as the manufacturing processes which create or add value to a product. The manufacturing processes can be classified as:

- Casting, foundry, or moulding processes
- Forming or metalworking processes
- Machining (material removal) processes
- Joining and assembly
- Surface treatments (finishing)
- Rapid prototyping
- Heat treating
- Other

Other manufacturing operations include inspection, testing, transportation, automated material handling and even packaging. In casting, the metal is heated sufficiently to make it into liquid and then poured into moulds of desired shapes. Various machining operations are turning, drilling and milling. Joining processes include welding, soldering, brazing and adhesive bonding. The process of heat treating is carried out to enhance various properties and include annealing and strengthening processes for metals and glasses. Surface processing includes cleaning, coating and thin film deposition, electroplating, anodising etc.

## Product Life Cycle and Life Cycle Cost

Manufacturing systems are dynamic and liable to change over time. Thus there is a traditional relation between a product's life cycle and the kind of manufacturing system used to make it. The life cycle consists of the following steps:

1) Start-up: New product or new company, low volumes

2) Rapid growth: Product becomes standardized in the market, higher volumes

3) Maturation: Designs become standard, process development becomes important

4) Commodity: Long life, standard of the industry type of product

5) Decline: Product replaced by improved products.

## Manufacturing System Design

A manufacturing system must consider two customers namely, the external that buys that the product and the internal that makes the product. The external customer may be global in scope, but the internal customer is critical in determining the design and manufacturing stages.

The complexity of the manufacturing system design where choices of system design trade off with parts variety.

## Basic Manufacturing Processes

### Machining

Tools used for machining are immobile power-driven units used to form or shape solid materials, specifically metals. The forming is done by removing extra materials from a work-piece. Machine tools make up the foundation of advanced industry and are utilized either indirectly or directly in the manufacturing of tool parts.

They are categorized under three main categories:

- Traditional Chip-making tools.
- Presses.
- Modern machine tools.

Traditional chip-making tools form the work-piece by trimming away the unwanted part accessible as chips. Presses implement a several shaping processes, which includes shearing, pressing, or elongating. Non-traditional machine tools implement light, electric powered, chemical, and sonic power; superheated gas; and high-energy compound beams to form the exotic supplies and materials that have been created to meet the requirements of modern technology.

## Joining

Every joining approach has particular design needs, while certain joint needs may propose a particular joining approach. Design for assembly, and fastener selection apply their own specifications.

Bolting is a standard fastening method, for instance, but welding may cut down the weight of assemblies. Naturally, joints intended for the two approaches would differ tremendously.

However, all joint patterns must consider features such as load factors, assembly effectiveness, operating surroundings, overhaul and upkeep, and the materials chosen.

Welding is generally a cost-effective approach to fabricate. It doesn't require overlapping materials, and so it removes excess weight brought on by other fastening methods. Fasteners don't have to be purchased and stored in stock. Welding also can minimize costs related to extra parts, for example angles mounted between parts.

## Forming

Metal forming is the approach of creating the metallic components by deforming the metal but not by removing, cutting, shredding or breaking any part. Bending, spinning, drawing, and stretching are a few important metal forming process in manufacturing. The metal press such as die and punching tools are implemented for this manufacturing process.

- Advantages:

  Same equipment can be utilized for manufacturing various components by simply changing the dies.

- Disadvantages:

  High apparatus and tooling expenses,

  Heat treatment must be applied afterwards.

## Casting

Casting is a manufacturing process in which a solid is dissolved into a liquid, heated to appropriate temperature (sometimes processed to change its chemical formula), and is then added into a mold or cavity. Thus, in just one step, complex or simple shapes can be crafted from any kind of metal that has the capability to be melted. The end product can have practically any arrangement the designer wants.

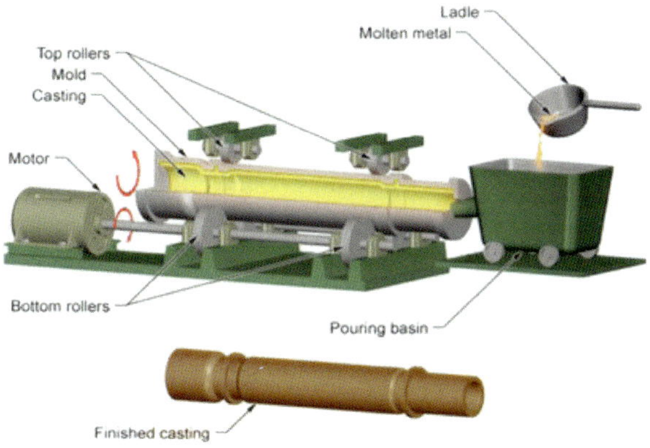


Furthermore, the reluctance to working challenges can be improved, directional attributes can be managed, and a pleasing look can be developed.

## Near Net Shape

Factories looking to maximize efficiency and keep costs low may choose a near net shape manufacturing strategy. Under a near net shape plan, the manufacturer aims to produce components that are as close to their finished state as possible. For example, a company that produces solid metal fasteners will simplify production to complete the fastener in as few steps as possible. Rather than start with a solid block of metal and use grinding and milling techniques to form and shape the fastener, the factory will likely create a die to cast the finished product in a single step.

Near net shape manufacturing can be applied to a wide variety of media. It's often used in metal working, particularly with titanium and other expensive alloys. This process can also be used in ceramics making, where spray or gel ceramics can offer a near net alternative to milling or machining. This type of production plan may also apply to thermoplastics production, where injection molding may replace more traditional manufacturing methods.

A factory wishing to instill a near net shape system may be required to overhaul the entire production process. Rather than an assembly-line process, net near requires combining many processes into a single step. This can greatly speed up the manufacturing process and increase production rates in the long run.

One of the primary advantages to near net shape production is the way it encourages an efficient use of material. Instead of slowly creating a widget out of a solid block of material, the firm uses exactly the correct amount of material to cast the widget in a mold or die. This process cuts waste, reduces energy use, and keeps costs low. It also reduces the total number of machines and tools needed to complete the process, and may reduce labor costs for some products.

Near net shape strategies also help companies create a virtually perfect product. Using a mold, companies eliminate errors and produce consistent finished goods. This strategy helps to enhance quality control throughout the firm.

This process is also associated with a number of disadvantages that should be considered before implementing near net shape plans. Near net often requires factories to invest in all new equipment, which can involve high upfront costs. This includes new metal dies or molds, as well as new tools and employee training. While this strategy often pays off in the form of lower long-term cost, the high initial investment may make near net out of reach for some companies.

## Metal Casting

Metal casting is the process of pouring molten metal into a shaped space so that it will cool and harden in that form. For many items, this process can be less expensive than machining the part out of a piece of solid metal. Even though the idea behind it seems singular and simple, there are many ways foundries cast objects. How they make a metal casting mold will depend on the metals used, the size of the run, and the shape, symmetry, and complexity of the casting.

### Metal Casting Processes

There are numerous metal casting processes implemented in the manufacturing of parts. Two main divisions can be identified by the fundamental nature of a mold they use. There is expendable casting and long-lasting mold casting. Expendable molds are utilized for a single metal casting while long-lasting molds are utilized for several. When considering manufacturing procedures, there are pros and cons to both of the processes.

### Expendable Mold

- Can only make one metal casting
- Made of sand, or other similar materials
- Binders used to support material hold its shape
- Mold that metal hardens in should be damaged to wipe out casting
- More complex geometries are feasible for casting

### Long-lasting Mold

- Can create many metal castings
- Generally made of metals or often a refractory ceramic
- It has parts that can close or open, permitting eradication of the casting
- Have to open mold limitations part designs

# Types of Metal Casting

## Centrifugal Casting

In this range of processes the melted material is required to disperse into the mold hole because of centrifugal speed. Centrifugal casting procedures can be categorized as true casting processes, which, when the melted material is added into the cavity, pushes it against the cavity walls where it stiffens into a hollow container; semi-centrifugal casting, varies from real centrifugal casting. In this process, the mold cavity is 100% filled with molten material so that the center, which is later on taken out, is exposed to low stress and is where inclusions and air is caught and the centrifuging approach in which melted material is poured into several mold cavities and which move around the core axis and shifted from the core axis and causes the cavities to be stuffed under excessive pressure.

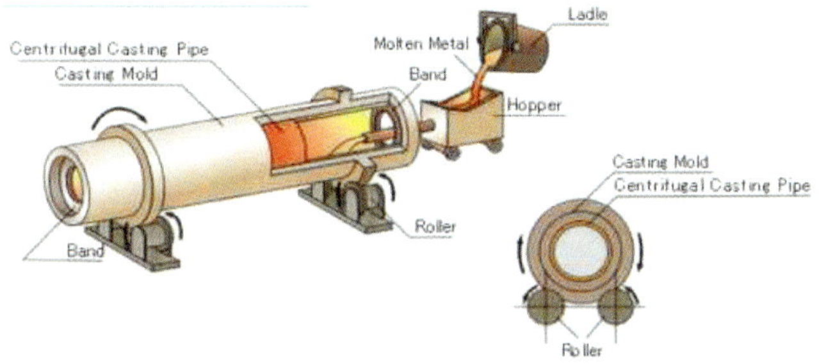


## Continuous Casting

In this process, a flow of molten material equates of a water-cooled hole and creates a continuous rod or strip, which is then chopped by a rounded saw. A new technique, generally known as circular continuous casting, entails the water-cooled hole (mold) oscillating as well as rotating at approximately 120 RPM during the casting procedure. Constant casting has an extremely high metal yield, around 98% in comparison with 87% in standard ingot-mold methods, outstanding quality of cast, manipulated grain sizes and the capability of casting unique cross-sectional designs.

## Die Casting

Metal die casting processes force the molten material into the hole of a steel cavity, termed as a die, within extremely high pressure, approximately 1000 – 30,000 psi. Categorization of die casting entails the kinds of machines used, the main types being used are hot-chamber and cold-chamber machines.

BASIC PRINCIPLE OF DIE CASTING

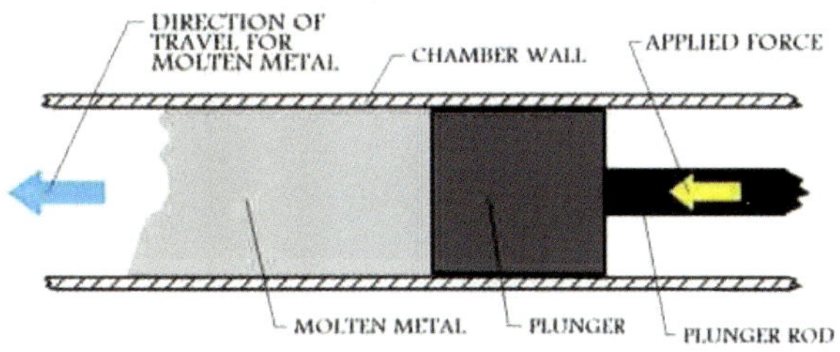

## Molding

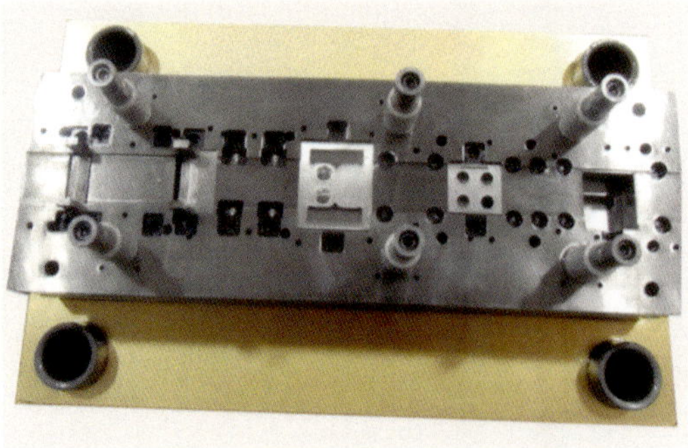

Molding or molding is the process of manufacturing by shaping liquid or pliable raw material using a rigid frame called a mold or matrix. This itself may have been made using a pattern or model of the final object.

A mold or mold is a hollowed-out block that is filled with a liquid or pliable material like plastic, glass, metal, or ceramic raw materials. The liquid hardens or sets inside the mold, adopting its shape. A mold is the counterpart to a cast. The very common bi-valve molding process uses two molds, one for each half of the object. Piece molding uses a number of different molds, each creating a section of a complicated object. This is generally only used for larger and more valuable objects.

The manufacturer who makes the molds is called the mold-maker. A release agent is typically used to make removal of the hardened/set substance from the mold easier. Typical uses for molded plastics include molded furniture, molded household goods, molded cases, and structural materials.

## Metal Forming

Metal forming is a general term for a large group, that includes a wide variety of manufacturing processes. Metal forming processes are characteristic in that the metal being processed is plastically deformed to shape it into a desired geometry. In order to plastically deform a metal, a force must be applied that will exceed the yield strength of the material. When low amounts of stress are applied to a metal it will change its geometry slightly, in correspondence to the force that is exerted. Basically it will compress, stretch, and/or bend a small amount. The magnitude of the amount will be directly proportional to the force applied. Also the material will return to its original geometry once the force is released. Think of stretching a rubber band, then releasing it, and having it go back to its original shape. This is called elastic deformation. Once the stress on a metal increases past a certain point, it no longer deforms elastically, but starts to undergo plastic deformation. In plastic deformation, the geometric change in the material is no longer directly proportional to stress and geometric changes remain after the stress is released; meaning that the material does not recover its shape. The actual level of stress applied to a metal where elastic deformation turns to plastic deformation is called the proportional limit, and is often difficult to determine exactly. The. 002 offset convention is usually used to determine the yield point, which is taken for practical purposes as the stress level where plastic deformation, (yielding), begins to occur.

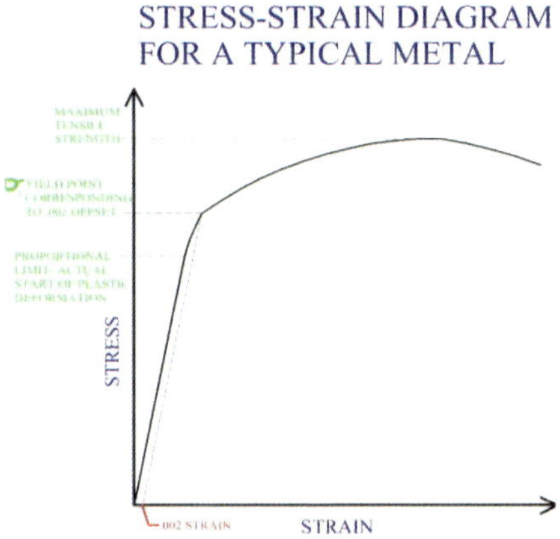


It can be seen by the stress-strain graph that once the yield point of a metal is reached and it is deforming plastically, higher levels of stress are needed to continue its deformation. The metal actually gets stronger, the more it is deformed plastically. This is called strain hardening or work hardening. As may be expected, strain hardening is a very important factor in metal forming processes. Strain hardening is often a problem that must be overcome, but many times strain hardening, when used correctly, is a vital part of the manufacturing process in the production of stronger parts.

## Flow Stress

During a metal forming operation, it is important to know the force and power that will be needed to accomplish the necessary deformation. The stress-strain graph shows us that the more a work piece is deformed plastically, the more stress is needed. The flow stress is the instantaneous value of the force necessary to continue the yielding and flow of the work material at any point during the process. Flow stress can be considered as a function of strain. The flow stress value can be used to analyze what is going on at any particular point in the metal forming process. The maximum flow stress may be a critical measurement in some metal forming operations, since it will specify the force and power requirements for the machinery to perform the process. The force needed at the maximum strain of the material would have to be calculated in order to determine maximum flow stress.

For different types of metal forming processes, the flow stress analysis may be different. For a process like forging, the maximum flow stress value would be very important. However, for a process like extrusion, where the metal is continuously being deformed and the different stages of the metal's deformation are occurring simultaneously, it is of interest to analyze the mean flow stress value.

## Strain Rate

The strain rate for any particular manufacturing metal forming process is directly related to the speed at which deformation is occurring. A greater rate of deformation of the work piece will mean a higher strain rate. The specific process and the physical action of the equipment being used has a lot to do with strain rate. Strain rate will affect the amount of flow stress. The effect strain rate has on flow stress is dependent upon the metal and the temperature at which the metal is formed. The strain rate with relation to flow stress of a typical metal at different temperatures is shown in figure.

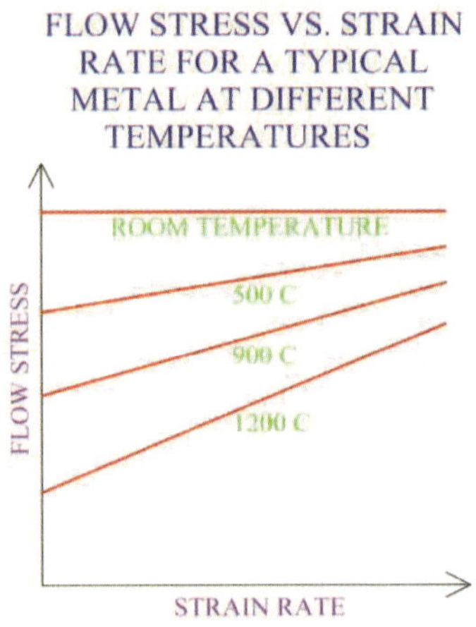


## Different Types of Metal Forming Processes

Metal forming processes can be classified under two major groups: Bulk deformation processes and sheet metal working processes. Bulk deformation is characteristic in that the work formed has a low surface area to volume ratio. In sheet metal working, the metal being processed will have a high surface area to volume ratio. The following is a brief overview of the major metal forming processes that will be covered in detail later.

## Bulk Deformation

Rolling: Rolling is a metal forming process that deforms the work by the use of rolls. Rolling processes include flat rolling, shape rolling, ring rolling, thread rolling, gear rolling, and the production of seamless tube and pipe by rotary tube piercing or roll piercing.

Forging: Forging is characteristic in the use of dies to compress and shape a work piece. The die may be flat or may contain an impression of a certain geometry.

Extrusion: Extrusion involves forming by forcing metal through a die opening, producing work of variable length and constant cross section.

Drawing: Drawing is similar to extrusion, in that a length of metal is made to flow through a die opening and forming is done over its cross section. The difference between drawing and extrusion is the application of force to the work piece. In extrusion the work is pushed through the die opening, in drawing the work is pulled through the die opening.

## Sheet Metal Working

Shearing: Shearing is the cutting of the work piece, this would include punching holes. Technically shearing does not involve shaping by plastic deformation, but it is a critical process in sheet metal working operations and should be understood along with metal forming processes.

Bending: Bending involves the deformation of the work by way of bending about a certain axis.

Deep Drawing: Deep drawing is a metal forming process in which a flat piece of plate or sheet is forced into a die cavity to take a shape, such as a cup.

## Effect of Temperature in Metal Forming

Properties of a metal change with an increase in temperature. Therefore, the metal will react differently to the same manufacturing operation if it is performed under different temperatures and the manufactured part may posses different properties. For these reasons, it is very important to understand the materials that we use in our manufacturing process. This involves knowing their behavior at various temperature ranges. In industrial metal forming manufacture, there are three basic temperature ranges at which the metal can be formed, cold working, warm working, and hot working.

## Cold Working

Cold working, (or cold forming), is a metal forming process that is carried out at room temperature or a little above it. In cold working, plastic deformation of the work causes strain hardening as discussed earlier. The yield point of a metal is also higher at the lower temperature range of cold forming. Hence, the force required to shape a part is greater in cold working than for warm working or hot working. At cold working temperatures, the ductility of a metal is limited, and only a certain amount of shape change may be produced. Surface preparation is important in cold forming. Fracture of the material can be a problem, limiting the amount of deformation possible. In fact, some metals will fracture from a small amount of cold forming and must be hot formed. One main disadvantage of this type of process is a decrease in the ductility of the part's material, but there are many advantages. The part will be stronger and harder due to strain hardening. Cold forming causes directional grain orientation, which can be controlled to produce desired directional strength properties. Also, work manufactured by cold forming can be created with more accurate geometric tolerances and a better surface finish. Since low temperature metal forming processes do not require the heating of the material, a large amount of energy can be saved and faster production is possible. Despite the higher force requirements, the total amount of energy expended is much lower in cold working than in hot working.

## Warm Working

Warm working, (or warm forming), is a metal forming process carried out above the temperature range of cold working, but below the recrystallization temperature of the metal. Warm working may be preferred over cold forming because it will reduce the force required to perform the operation. Also, the amount of annealing of the material that may have been necessary for the cold formed part may be less for warm working.

## Hot Working

Hot working, (or hot forming), is a metal forming process that is carried out at a temperature range that is higher than the recrystallization temperature of the metal being formed. The behavior of the metal is significantly altered, due to the fact that it is above its recrystallization temperature. Utilization of different qualities of the metal at this temperature is the characteristic of hot working.

Although many of these qualities continue to increase with increasing temperature, there are limiting factors that make overly high temperatures undesirable. During most metal forming processes the die is often cold or slightly heated. However, the metal stock for hot working will usually be at a higher temperature relative to the die. In the design of metal forming process, it is critical to consider the flow of metal during the forming of the work. Specific metal flow, for different forming processes, is discussed in latter sections under each specific process. For metal forming manufacturing, in general, the temperature gradient between the die and the work has a large effect on metal flow during the process. The metal nearer to the die surfaces will be cooler than the metal closer to the inside of the part, and cooler metal does not flow as easily. High temperature gradients, within the work, will cause greater differences in flow characteristics of different sections of the metal, these could be problematic. For example, metal flowing significantly faster at the center of the work compared to cooler metal near the die surfaces that is flowing slower, can cause part defects. Higher temperatures are harder to maintain throughout the metal forming process. Work cooling during the process can also result in more metal flow variations. Another consideration with hot forming manufacture, with regard to the temperature at which to form the part, is that the higher the temperature the more reactive the metal is likely to be. Also if a part for a hot working process is too hot then friction, caused during the process, may further increase heat to certain areas causing melting, (not good), in localized sections of the work. In an industrial hot metal working operation, the optimum temperature should be determined according to the material and the specific manufacturing process.

When above its recrystallization temperature a metal has a reduced yield strength, also no strain hardening will occur as the material is plastically deformed. Shaping a metal at the hot working temperature range requires much less force and power than in cold working. Above its recrystallization temperature, a metal also possesses far greater ductility than at its cold worked temperature. The much greater ductility allows for massive shape changes that would not be possible in cold worked parts. The ability to perform these massive shape changes is a very important characteristic of these high temperature metal forming processes.

The work metal will recrystallize, after the process, as the part cools. In general, hot metal forming will close up vacancies and porosity in the metal, break up inclusions and eliminate them by distributing their material throughout the work piece, destroy old weaker cast grain structures and produce a wrought isotropic grain structure in the part. These high temperature forming processes do not strain harden or reduce the ductility of the formed material. Strain hardening of a part may or may not be wanted, depending upon the application. Qualities of hot forming that are considered disadvantageous are poorer surface finish, increased scale and oxides, decarburization, (steels), lower dimensional accuracy, and the need to heat parts. The heating of parts reduces tool life, results in a lower productivity, and a higher energy requirement than in cold working.

## Selection of Temperature Range for a Metal Forming Operation

Production at each of these temperature ranges has a different set of advantages and disadvantages. Sometimes, qualities that may be undesirable to one process may be desirable to another. Also, many times work will go through several processes. The goal is to design the manufacture of a part in such a way as to best utilize the different qualities to meet or enhance the specifications of the part. To produce a strong part with excellent surface finish, then a cold forming process could be a good choice. However, to produce a part with a high ductility a hot forming process may be best. Sometimes the advantages of both hot forming and cold forming are utilized when a part is manufactured by a series of processes. For example, hot working operations may first be performed on a work piece to achieve large amounts of shape change that would not be possible with cold forming due to strain hardening and limited ductility. Then the last process that completes the manufacture of the part is a cold working operation. This process does not require a significant shape change, since most of the deformation was accomplished by the hot forming process. Having a cold forming process last will finish the shape change, while strengthening the part, giving a good surface finish and highly accurate tolerances.

### Friction and Lubrication in Metal Forming

Metal forming processes are characteristic of high pressures between two contacting surfaces. In hot forming operations, these high pressures are accompanied by extreme temperatures. Friction and die wear are a serious consideration in metal forming manufacturing. A certain amount of friction will be necessary for some metal forming processes, but excessive friction is always undesirable. Friction increases the amount of force required to perform an operation, causes wear on tooling, and can affect metal flow, creating defects in the work.

Where friction is involved, lubricants can usually help. For some metal forming processes and materials no lubrication is used, but for many lubrication is applied to contacting surfaces to reduce friction forces. Lubricants used in industry are different depending upon the type of metal forming process, the temperature at which the operation occurs, and the type of material formed. Lubricants should be effective and not produce any toxic fumes. Lubricants used in manufacturing industry for metal forming processes include, vegetable and mineral oils, soaps, graphite dispensed in grease, water based solutions, solid polymers, wax, and molten glass.

## Machining Processes

Machining is a term that covers a large collection of manufacturing processes designed to remove unwanted material, usually in the form of chips, from a workpiece. Machining is used to convert castings, forgings, or preformed blocks of metal into desired shapes, with size and finish specified to fulfill design requirements. Almost every manufactured product has components that require machining, often to great precision. Therefore, this collection of processes is one of the most important of the basic manufacturing processes because of the value added to the final product. By the same token, machining processes are often the most expensive. The majority of industrial applications of machining are in metals. Although the metal cutting process has resisted theoretical analysis because of its complexity, the application of these processes in the industrial world is widespread. Machining processes are performed on a wide variety of machine tools. Figure shows an example of a machine tool-a dual-turret numerically controlled (NC) lathe. Workpieces are held in workholding devices, such as a three-jaw chuck. The tools used to cut metal are in the turrets. Other examples of basic machine tools are: Milling machines, drill presses, grinders, shapers, broaching machines, and saws. Each of the basic machine tool types has many different configurations. Lathes, for example, may be engine lathes, turret lathes, tracer lathes, or automatic-screw machines. Lathes have followed the trend of other machine tools, and NC lathes can now be routinely purchased. The primary chip formation processes are listed below, with alternative versions in parentheses. Each process is performed on one or more of the basic machine tools. For example, drilling can be performed on drill presses, milling machines, lathes, and some boring machines: Turning (boring, facing, cutoff, taper turning, form cutting, chamfering, recessing, thread cutting). Shaping (planning, vertical shaping) Milling (hobbing, generating, thread milling) Drilling (reaming, tapping, spot facing, counter boring, countersinking) Sawing (filing) Abrasive machining (grinding, honing, lapping) Broaching (internal and surface) Processes can be combined into multiple-capability machines, known as machining centers. The machining center is capable of performing the machining processes normally performed on a milling machine, drilling machine, and a boring mill and is numerically controlled. The position and velocity of the tool with respect to the work is under feedback control. Different tools can be automatically inserted into the spindle as needed to do different machining processes. The horizontal spindle machine was one of the first NC machining centers to be able to change workpiece pallets. For each of the basic machine tool types, there are many different kinds of workholders, cutting tools, and cutting tool holders, resulting in a rather formidable list of equipment and processes.

## Joining

Joining comprises a large number of processes used to assemble individual parts into a larger, more complex component or assembly. The individual parts of a component meet at the joints. Joints transmit or distribute forces generated during service from one part to the other parts of the assembly. A joint can be either temporary or permanent. The five joint types that are predominately used in the joining of parts are the butt, tee, corner, lap, and edge joints.

The selection of an appropriate design to join parts is based on several considerations related to both the product and the joining process. Product related considerations include codes and standards, fitness for service, aesthetics, manufacturability, reparability, reliability, inspectability, safety, and unit cost of fabrication. Joining process considerations include material types and thicknesses, joint geometry, joint location and accessibility, handling, jigging and fixturing, distortion control, productivity, and initial and recurring manufacturing costs. Additional considerations include whether the joint is fabricated in a shop or at a remote site, possibilities for failure, and containment in case of a catastrophic failure (e.g., a nuclear reactor vessel).

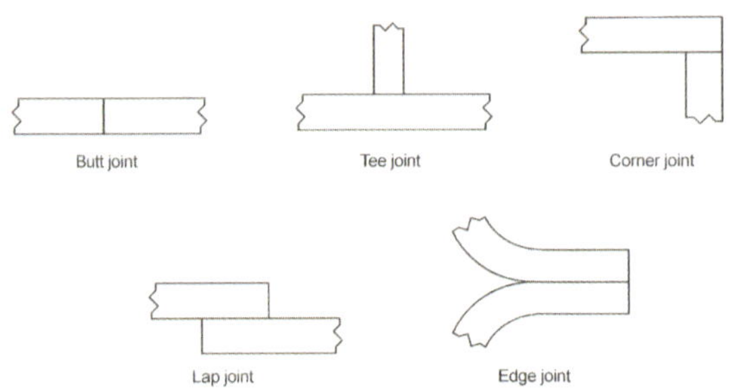


Figure: Types of joints.

The design or selection of appropriate joint type is determined primarily by the type of service loading the assembly will be exposed to during its service life. For example, in metallic structures, butt joints are preferred over tee, corner, lap, or edge joints for components subjected to fatigue loading, while a lap joint would be optimum for an adhesive- bonded structure. However, since high-strength adhesives are typically weak in peel, the lap joint should be loaded in shear. To effectively transfer loads through the adhesive, the substrates (or adherends) are over- lapped so that the adhesive is loaded in shear. The specific joint design aspects, such as the size, length, and relative orientation of the joint, are based on stress calculations that are results of the anticipated service loads, properties of materials, properties of sections, and appropriate structural design requirements. An ideal joint is one that effectively transmits forces among the joint members and throughout the assembly, meets all structural design requirements, and can still be produced at a minimal cost. This involves selection and application of good design practices based on a thorough understanding of the available joining processes.

Assembly imposes constraints on the design. The parts must be de- signed so that they not only can be assembled and joined together to provide the needed function but also are easy to handle, insert, retain, and verify that they have been assembled correctly. Because assembly is an integrative process, problems with detail part designs often surface when they are assembled. For example, parts may not fit together properly, tools may not reach in the space provided, and parts may be incorrectly assembled. These problems often require extensive rework, resulting in costly schedule slippage and undesirable design compromises.

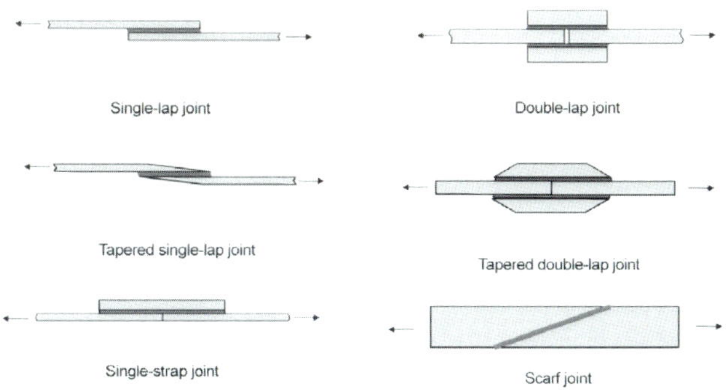


Figure: Typical adhesive-bonded joint configurations. Note that the adhesive is loaded in shear in all configurations

The importance of assembly as a design constraint has resulted in a greatly increased emphasis on assembly in the design process. Design and manufacturing practice now focuses on ensuring that parts conform to specifications and that variability and randomness are minimized, and on making non-value-added operations such as orienting and handling as simple and easy to perform as possible. Many product design departments now improve the ease with which products are assembled by using design for assembly (DFA) techniques, which seek to ensure ease of assembly by developing designs that are easy to assemble.

The aim of DFA is to simplify the product so that the cost of assembly is reduced. However, DFA techniques often result in improved quality and reliability, along with a reduction in production equipment and part in- ventory. These secondary benefits often outweigh the cost reductions in assembly.

## General Guidelines for DFA

- Minimize part count by incorporating multiple functions into single parts.
- Modularize multiple parts into single subassemblies.
- Assemble in open space, not in confined spaces; never bury important components.

- Make parts such that it is easy to identify how they should be oriented for insertion.
- Prefer self-locating parts.
- Standardize to reduce part variety.
- Maximize part symmetry.
- Eliminate parts that will tangle.
- Color code parts that are different but shaped similarly.
- Prevent nesting of parts; prefer stacked assemblies.
- Provide orienting features on nonsymmetries.
- Design the mating features for easy insertion.
- Provide alignment features.
- Insert new parts into an assembly from above.
- Eliminate reorientation of both parts and assemblies.
- Eliminate fasteners.
- Place fasteners away from obstructions; design in fastener access.
- Deep channels should be sufficiently wide to provide access to fastening tools; eliminate channels if possible.
- Provide flats for uniform fastening and fastening ease.
- Ensure sufficient space between fasteners and other features for a fastening tool
- Prefer easily handled parts.

Generally, a concept design is developed and then evaluated against each of these guidelines. Design modifications are then made to satisfy the guideline. There is no guarantee that a given guideline will apply to a particular design problem. Many of these guidelines are similar or the same as the rules of concurrent engineering.

## Concurrent Engineering Rules

- Ensure that parts most likely to require maintenance are easily accessible.
- Ensure that the degree of maintenance of your product is consistent
- With your company's policy on making, stocking, and supplying spare parts.
- Ensure tools needed for installation and maintenance are as inexpensive and common as possible.

- The decisions made in the first 15% of a product development process fix 85% of the downstream quality and cost of the product.
- Include all experts actively.
  - Resist making irreversible decisions.
  - Continually optimize the designed product and the design process.
  - Prefer concepts that are easy to manufacture.
  - Prefer concepts that are easy to assemble.
  - Integrate design and manufacturing.
  - Do not over constrain or under constrain the design.
  - Look ahead of the current state of the design to anticipate problems.
  - Reduce the number of parts.
  - Increase interchangeability of parts; standardize parts; minimize variation in parts.
  - Modularize functions and subassemblies.
  - Design multifunctional and multiple-use parts.
  - Avoid flexible components.
  - Avoid separate fasteners.
  - Improve robustness.
  - Allocate time/man power based on cost-benefit analysis of a proposed action.
  - Maximize yield of existing equipment.
  - Keep assemblies/components as independent as possible.
  - Maximize tolerances.
  - Test only what can be quantified; actively search for testable aspects of a design.
  - Minimize machining setups and reorientations.
  - Design parts for feeding and insertion into machines.
  - Perform functional analysis.
  - Tailor the manufacturing process to the character of the product.
  - Study producibility and usability.
  - Design the fabrication process.

- Design the assembly sequence for top-down assembly.
- Minimize assembly instructions.
- Use known/proven vendors and suppliers.
- Use new technologies only when necessary.
- Identify subassemblies as soon as possible in the design process.
- Do engineering changes in batches.
- Integrate quality control with assembly.
- Match assembly processes to tolerances.
- Operate on a minimum inventory.

While some of these guidelines and rules may not always be applicable, and even occasionally contradict each other, they are certainly worth considering during the design process.

## Types of Joining

Joining processes include welding, brazing, soldering, mechanical fastening, and adhesive bonding. Mechanical fastening can be used to provide either temporary or permanent joints, while adhesive bonding, welding, brazing, and soldering processes are mainly used to provide permanent joints. Mechanical fastening and adhesive bonding usually do not cause metallurgical reactions. Consequently, these methods are often preferred when joining dissimilar combinations of materials and for joining polymer-matrix composites that are sensitive to extreme heat. Welding processes are divided into two broad classes: fusion welding and solid- state welding.

## Fusion Welding

Fusion welding processes involve localized melting and solidification and are normally used when joining similar material combinations or materials belonging to the same family (e.g., joining one type of stainless steel with another type). In fusion welding, the weld can be made by simply melting the edges of the two workpieces and allowing them to fuse together on cooling. This type of weld is referred to as an autogenous weld. The other method is to add extra material during the welding process through the melting of an electrode or filler wire during the welding process. In both cases, the welded area will have a microstructure and properties that are different from the parent metal. The three predominant zones in a fusion weld are the fusion zone, a heat-affected zone (HAZ), and the base metal as shown in figure. The weld deposit itself will have a cast structure of often a complex composition. Between the weld deposit and the parent metal is an HAZ that did not melt during welding but reached very high temperatures. Grain growth due to the high temperatures is commonly encountered in the HAZ.

The types of welds commonly used with fusion welding processes are shown in Fig. Joint designs and clearances that overwhelmingly trap the beam energy within the joint cavity are preferred for increased process efficiency. When joining thick sections, the preferred joint designs allow the weld metal to freely shrink without causing cracking. In addition, multipass welding is used for thick sections to provide for full penetration. The distinctive microstructure of a multipass weld compared to a single- pass weld is shown in Fig. To maintain tolerances, distortion due to localized heating and cooling must be prevented or controlled during welding by the use of jigs and fixtures. Residual stresses can lead to distortion after welding and are often minimized by a stress-relief anneal after welding.

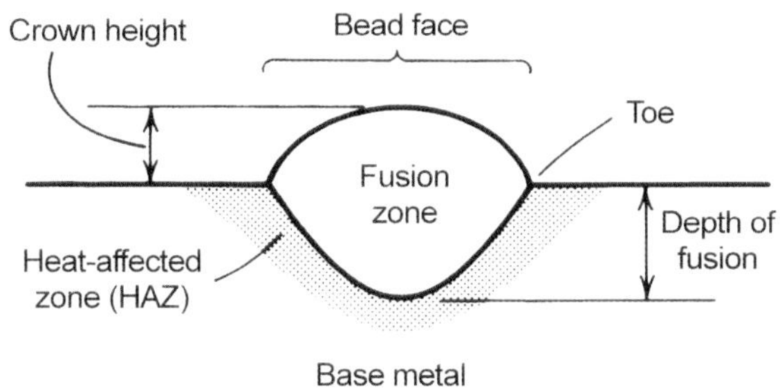


Figure: Weld bead geometry showing fusion zone, heat-affected zone, and base metal

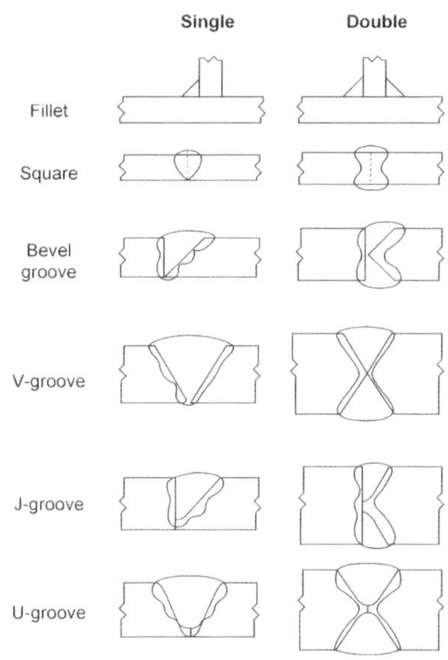


Figure: Types of welds

Thermal Welding: Heat is provided by an oxyfuel gas flame, mostly for manual welding, or by a thermit reaction for joining heavy sections such as rails. Portability is a great advantage.

Electric Arc Welding: Electric arc welding processes use electricity to produce the intense heat necessary for welding. Some electric arc processes use a consumable electrode that melts and becomes part of the weld metal that is deposited, whereas others use a nonconsumable electrode that does not melt and does not become part of the weld deposit. Consumable electrode welding uses a filler rod as the electrode. Atmospheric protection of the molten weld metal is provided by a slag produced by the filler rod or by an externally supplied inert gas. Flux-cored wire allows a continuous and mechanized operation. The flux is supplied as a powder in submerged arc welding for horizontal welds, and a resistive slag pool protects the weld zone in electroslag welding of thick plates. Inert gas provides the protection in some welding processes, such as gas metal arc welding (GMAW) and gas tungsten arc welding (GTAW).

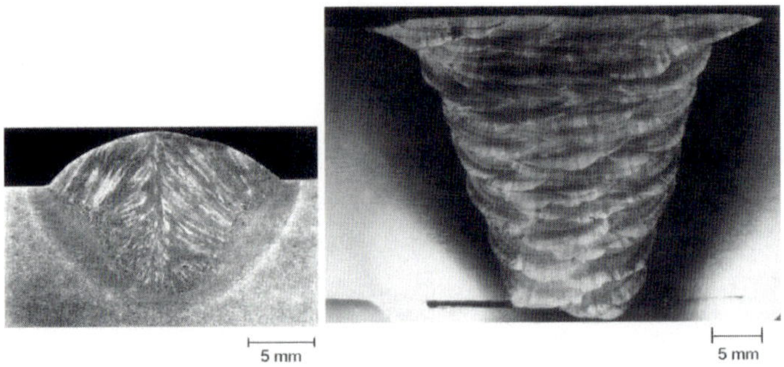


Figure: Weld microstructures showing the fusion zone, heat-affected zone, and base metal for (a) single-pass bead-on-plate weld in A-710 steel and (b) multipass weld in 304 stainless steel.

Figure: Welding using shielded metal arc welding process

Resistance Welding: After the two parts have been pressed together, electric current passes through the joint to heat and melt the interface. Pressure is kept on until solidification of the melt is complete. Spot welding is widely employed in building automotive bodies using welding robots. Seam welding, a continuous stream of spot welds, is used for making beams and box sections.

High-Energy Beam Welding: Highly concentrated beams of electrons impinge on the weld zone in electron beam (EB) welding. When the work- piece is enclosed with the gun in a vacuum chamber, a high vacuum protects the surfaces but increases cost and lowers production rates. Out-of- chamber welding is also possible. Gas ($CO_2$) or solid-state (Nd-YAG) lasers have seen increasing application for joining not only difficult metals and delicate parts but also sheets of different thicknesses for tailored blanks used in automobile body construction. All high-energy beam processes have the advantage that the HAZ is small, as shown in the EB weld microstructure in figure.

## Solid-state Welding

Since solid-state welding processes do not involve melting and solidification, they are often suitable for joining not only similar but also dissimilar materials. Solid-state welding processes also have special joint design or part cross-section requirements. For example, continuous drive and inertia friction welding processes require that one of the parts exhibit a circular or near-circular cross section.

Diffusion bonding is a solid-state welding process that allows joining of a variety of structural materials, both metals and nonmetals. However, diffusion bonding requires an extremely smooth surface finish to provide intimate surface contact, a high temperature, and a high pressure, first to allow intimate contact of the parts along the bond interface, followed by plastic deformation of the microscopic surface asperities, and then to promote diffusion across the bond interface. The need to apply pressure while maintaining part alignment imposes a severe limitation on joint design. When perfectly clean surfaces are brought into intimate contact, interatomic bonds form a joint. Bond strength is greatest when the mating metals are mutually soluble, but good bonds can be obtained with dissimilar, not otherwise weldable metals and with highly differing thicknesses. When an exceptional surface finish is difficult to achieve, a metallurgically compatible, low-melting interlayer can be inserted between the parts to produce a transient liquid phase on heating. On subsequent cooling this li6uid phase undergoes progressive solidification, aided by diffusion across the solid/liquid interfaces, and thereby joins the parts. This process is similar to a brazing process.

Cold Welding: Sufficient pressure must be exerted to establish conformance of surfaces. Sliding and deformation accompanied by surface expansion are needed to break up oxides and other adsorbed films. Cold welding processes differ only in the method of providing these conditions. Complex tubular parts such as refrigerator evaporator plates can be made by depositing a parting agent in a pattern to prevent welding. After roll bonding the passages may be inflated.

Forge Welding: As a generic term, forge welding applies to bonding by deformation at the hot-working temperature. Large surface extension in hot-roll bonding creates strong bonds for cladding, as in bonding copper- nickel surface layers to a copper core for some U.S. coins.

Friction Welding: Heat is produced by friction between a rotating and stationary part; again, some melt may form that is expelled together with oxidized metal. Localization of heat allows welding of dissimilar metals and of very different dimensions (e.g., a thin stem to a large head for an internal combustion engine valve).

From a metallurgical perspective, the application of both fusion welding and solid-state welding processes must be evaluated using appropriate weldability test methods for their ability to either recreate or retain base metal characteristics across the joint. When metallurgical reactions occur, they can either benefit or adversely affect the properties of the joint. These weldability evaluations need to combine material, process, and procedure aspects to identify combinations that would provide a weld joint with an acceptable set of properties.

## Brazing

Brazing is a process for joining solid metals in close proximity by introducing a li6uid metal that melts above 430°C (840°F). A sound brazed joint generally results when an appropriate filler alloy is selected, the parent metal surfaces are clean and remain clean during heating to the flow temperature of the brazing alloy, and a suitable joint design is used that allows capillary action. Strong, uniform, leak-proof joints can be made rapidly, inexpensively, and even simultaneously. Joints that are inaccessible and parts that may not be joinable at all by other methods can often be joined by brazing. Complicated assemblies comprising thick and thin sections, odd shapes, and differing wrought and cast alloys can be turned into integral components by a single trip through a brazing furnace or a dip pot. Metal as thin as 0.01 mm (0.0004 in.) and as thick as 130 mm (6 in.) can be brazed.

Brazed joint strength is high. The nature of the interatomic (metallic) bond is such that even a simple joint, when properly designed and made, will have strength equal to or greater than that of the parent metal. The fact that brazing does not involve any substantial melting of the base metals offers several advantages over other welding processes. It is generally possible to maintain closer assembly tolerances and to produce a cosmetically neater joint without costly secondary operations. Even more important is that brazing can join dissimilar metals (or metals to ceramics) that, because of metallurgical incompatibilities, cannot be joined by traditional fusion welding processes. If the base metals do not have to be melted to be joined, it does not matter that they have widely different melting points. Therefore, steel can be brazed to copper as easily as to another steel. Brazing also generally produces less thermally induced distortion, or warping, than does fusion welding. An entire part can be brought up to the same brazing temperature, thereby preventing the kind of localized heating that causes distortion in welding.

Finally, and perhaps most important to the manufacturing engineer, brazing readily lends itself to mass production techniques. It is relatively easy to automate because the application of heat does not have to be localized, as in fusion welding, and the application of filler metal is less critical. In fact, given the proper clearance conditions and heat, a brazed joint is not dependent on operator skill, as are most fusion welding processes. Automation is also simplified by the fact that heat can be applied to the joint by many means, including torches, furnaces, induction coils, electrical resistance, and dipping. Several joints in one assembly often can be produced in one multiple-braze operation during one heating cycle, further enhancing production automation.

### Soldering

Soldiering is a joining process by which two substrates are joined together using a filler metal (solder) with a liquidus ≤ 450°C (≤840°F). The substrate materials remain solid during the bonding process. The solder is usually distributed between the properly fitted surfaces of the joint by capillary action.

## Subtractive Manufacturing

Subtractive manufacturing is a process by which 3D objects are constructed by successively cutting material away from a solid block of material. Subtractive manufacturing can be done by manually cutting the material but is most typically done with a CNC Machine. Advanced CNC machines utilize multiple tools and cut around at least three (x, y, and z) axes such that they minimize the requirement for designers to flip the block. One of the principal advantages to subtractive manufacturing is the ability to machine an extremely thin piece of plastic into a living hinge. This kind of process is simply not yet possible in a 3D printer. For those prototypes that require living hinge components it is useful to produce certain parts using additive manufacturing while using the CNC machine for specialty components like a living hinge.

### Advantages of Subtractive Manufacturing and Prototyping

Besides eliminating the layering of additive processes, subtractive processes have other advantages. They offer a variety of surface finishes, doing away with the “stepped” surfaces often found in many additive processes. The finish can be functionally important if parts must slide and cosmetically important if the prototypes are to be used in market testing.

However, because the process is removing material instead of adding it, milling undercuts can sometimes be difficult. Machining also tends to be somewhat more expensive than the additive processes.

One long viewed disadvantage of subtractive manufacturing has been the programming required to turn CAD instructions and information on an object into machine executable code. This disadvantage no longer exists. The automation of tool path generation is the key that makes subtractive prototyping competitive with the additive methods. Today's vendors have found ways to make CNC tool-path generation easy and efficient. Proto Labs, for example, has found a way to automatically translate a 3D CAD design into instructions for high-speed CNC milling and uses a large cluster of computers to quickly generate code. With the Roland machines, you export your CAD model as an. STL. DXF. 3DM or. IGS/IGES file. Through Roland's SRP Player software, you use its wizard to process your data. Based on data you provide the wizard, it will handle feeds and speeds, cut depth, tool selection and G-Code programming.

Subtractive RP closely mimics the surface finishes and tolerances of injection-molded parts, enabling you to test the design for function. Another development in subtractive RP is the geometry it can now support. The tool path software can handle more complex shapes, often on par with additive processes. Additive is still the choice for certain geometries, such as a sphere within a sphere or parts that interlock in unusual ways.

Thus, the advantages of subtractive prototyping include a wide selection of end-use materials, good dimensional control and surface finish, and a high degree of repeatability suitable for end-use manufacture. The drawbacks are that there is some material waste, and geometry limitations.

## Rapid Injection Molding

Sometimes viewed as a subset of subtractive prototyping, rapid injection molding (RIM) is done by injecting thermoplastic resins into a mold, just as is done in production injection molding. What makes the process "rapid" is the technology used to produce the mold, which is often made from aluminum instead of the traditional steel used in production molds.

Molded parts are strong and can have excellent finishes. It is also the industry standard production process for plastic parts, so there are inherent advantages to prototyping in the same process if the situation allows. Almost any engineering grade resin can be used, so you are not constrained by the material limitations of the prototyping process.

There is an initial tooling cost with RIM that does not occur with any of the additive processes or with CNC machining. In most cases, though, it makes sense to do one or two rounds of rapid prototypes (subtractive or additive) to check fit and function before moving to injection molding. ProtoMold is one of the suppliers of RIM technology.

A number of service bureaus can meet your subtractive and RIM needs. Some do the work in their facilities. Some contract the work out to others.

# Additive Manufacturing

Additive manufacturing, also known as 3D printing, is a process that creates a physical object from a digital design.

## Additive Manufacturing Processes

There are a variety of different additive manufacturing processes:

### Material Extrusion

Material extrusion is one of the most well-known additive manufacturing processes. Spooled polymers are extruded, or drawn through a heated nozzle mounted on a movable arm. The nozzle moves horizontally while the bed moves vertically, allowing the melted material to be built layer after layer. Proper adhesion between layers occurs through precise temperature control or the use of chemical bonding agents.

### Directed Energy Deposition

The process of directed energy deposition (DED) is similar to material extrusion, although it can be used with a wider variety of materials, including polymers, ceramics and metals. An electron beam gun or laser mounted on a four- or five-axis arm melts either wire or filament feedstock or powder.

### Material Jetting

With material jetting, a print head moves back and forth, much like the head on a 2D inkjet printer. However, it typically moves on x-, y- and z-axes to create 3D objects. Layers harden as they cool or are cured by ultraviolet light.

### Binder Jetting

The binder jetting process is similar to material jetting, except that the print head lays down alternate layers of powdered material and a liquid binder.

### Sheet Lamination

Laminated object manufacturing (LOM) and ultrasonic additive manufacturing (UAM) are two sheet lamination methods. LOM uses alternate layers of paper and adhesive, while UAM employs thin metal sheets conjoined through ultrasonic welding. LOM excels at creating objects ideal for visual or aesthetic modeling. UAM is a relatively low-temperature, low-energy process used with various metals, including titanium, stainless steel and aluminum.

### Vat Polymerization

With vat photopolymerization, an object is created in a vat of a liquid resin photopolymer. A process called photopolymerization cures each microfine resin layer using ultraviolet (UV) light precisely directed by mirrors.

## Powder Bed Fusion

Powder bed fusion (PBF) technology is used in a variety of AM processes, including direct metal laser sintering (DMLS), selective laser sintering (SLS), selective heat sintering (SHS), electron beam melting (EBM) and direct metal laser melting (DMLM). These systems use lasers, electron beams or thermal print heads to melt or partially melt ultra-fine layers of material in a three-dimensional space. As the process concludes, excess powder is blasted away from the object.

## Additive Manufacturing Technologies

Sintering is the process of creating a solid mass using heat without liquefying it. Sintering is similar to traditional 2D photocopying, where toner is selectively melted to form an image on paper.

Within DMLS, a laser sinters each layer of metal powder so that the metal particles adhere to one another. DMLS machines produce high-resolution objects with desirable surface features and required mechanical properties. With SLS, a laser sinters thermoplastic powders to cause particles to adhere to one another.

By contrast, materials are fully melted in the DMLM and EBM processes. With DMLM, a laser completely melts each layer of metal powder while EBM uses high-power electron beams to melt the metal powder. Both technologies are ideal for manufacturing dense, non-porous objects.

Stereolithography (SLA) uses photopolymerization to print ceramic objects. The process employs a UV laser selectively fired into a vat of photopolymer resin. The UV-curable resins produce torque-resistant parts that can withstand extreme temperatures.

## Additive Manufacturing Materials

It is possible to use many different materials to create 3D-printed objects. AM technology fabricates jet engine parts from advanced metal alloys, and it also creates chocolate treats and other food items.

## Thermoplastics

Thermoplastic polymers remain the most popular class of additive manufacturing materials. Acrylonitrile butadiene styrene (ABS), polylactic acid (PLA) and polycarbonate (PC) each offer distinct advantages in different applications. Water-soluble polyvinyl alcohol (PVA) is typically used to create temporary support structures, which are later dissolved away.

## Metals

Many different metals and metal alloys are used in additive manufacturing, from precious metals like gold and silver to strategic metals like stainless steel and titanium.

## Ceramics

A variety of ceramics have also been used in additive manufacturing, including zirconia, alumina and tricalcium phosphate. Also, alternate layers of powdered glass and adhesive are baked together to create entirely new classes of glass products.

## Biochemicals

Biochemical healthcare applications include the use of hardened material from silicon, calcium phosphate and zinc to support bone structures as new bone growth occurs. Researchers are also exploring the use of bio-inks fabricated from stem cells to form everything from blood vessels to bladders and beyond.

## Additive Manufacturing Applications

Additive manufacturing is already used to produce an impressive array of products - everything from food creations to jet engine parts.

## Aerospace

AM excels at producing parts with weight-saving, complex geometric designs. Therefore, it is often the perfect solution for creating tight, strong aerospace parts.

In August 2013, NASA successfully tested an SLM-printed rocket injector during a hot fire test that generated 20,000 pounds of thrust. In 2015, the FAA cleared the first 3D-printed part for use in a commercial jet engine. CFM's LEAP engine features 19 3D-printed fuel nozzles. At the 2017 Paris Air Show, FAA-certified, Boeing 787 structural parts fabricated from titanium wire were displayed, according to Aviation Week.

## Automotive

CNN reported that the McLaren racing team is using 3D-printed parts in its Formula 1 race cars. A rear wing replacement took about 10 days to produce instead of five weeks. The team has already produced more than 50 different parts using additive manufacturing. In the auto industry, AM's rapid prototyping potential garners serious interest as production parts are appearing. For example, aluminum alloys are used to produce exhaust pipes and pump parts, and polymers are used to produce bumpers.

## Healthcare

At the New York University School of Medicine, a clinical study of 300 patients will evaluate the efficacy of patient-specific, multi-colored kidney cancer models using additive manufacturing. The study will examine whether such models effectively assist surgeons with pre-operative assessments and guidance during operations.

Global medical device manufacturing company Stryker are funding a research project in Australia that will use additive manufacturing technology to create custom, on-demand 3D printed surgical implants for patients suffering from bone cancer.

In general, healthcare applications for additive manufacturing are expanding, particularly as the safety and efficacy of AM-built medical devices is established. The fabrication of one-of-a-kind synthetic organs also shows promise.

### Product Development

As the potential for AM's design flexibility is realized, once impossible design concepts are now being successfully re-imagined. Additive manufacturing unleashes the creative potential of designers who can now operate free of the constraints under which they once labored.

### Advantages of Additive Manufacturing

Additive manufacturing allows the creation of lighter, more complex designs that are too difficult or too expensive to build using traditional dies, molds, milling and machining.

AM also excels at rapid prototyping. Since the digital-to-digital process eliminates traditional intermediate steps, it is possible to make alterations on the run. When compared to the relative tedium of traditional prototyping, AM offers a more dynamic, design-driven process.

Whether additive manufacturing is used for prototyping or production, lead times are frequently reduced. Lead times for certain jet engine parts have been reduced by a year or more. Also, parts once created from multiple assembled pieces are now fabricated as a single, assembly-free object.

In designing everything from bridges to skyscrapers, engineers have long sought to minimize weight while maximizing strength. With additive manufacturing, designers realize the dream of utilizing organic structures to greatly reduce the weight of objects. For example, in GE's GrabCAD® Bracket Challenge, the winning design was just as strong as the original bracket, even though it weighed almost 84 percent less.

## Rapid Manufacturing

Rapid manufacturing, also known as direct manufacturing/direct fabrication/ digital manufacturing, has been defined in various ways. One widely accepted definition is "the use of an additive manufacturing process to construct parts that are used directly as finished products or components". If this process occurs in the R&D stage, it is called rapid prototyping (RP). In the USA, the term "Solid Freeform Fabrication" is preferred to rapid prototyping or RM. According to Plesseria et al., starting with a 3-D CAD part model, this technique converts the model to a series of layers using software; these layers are transferred to the building machine in which they are "printed" from the material by different processes. After printing one layer, a new layer of material is deposited and so on. Postprocessing treatment may be supplemented. In other words, RM is the fabrication of parts or components using additive manufacturing technologies. The part is shaped layer- by-layer and could be used as a functional product. In this technique, the removal cutting tools are not required to produce the physical products and the parts with complicated geometry may be fabricated. RM is sometimes considered one of the RP applications. However, there is no clear distinction among definitions. RM's outputs are often usable products. Stereolithography (SLA), laser sintering (LS), and fused deposition modeling (FDM) are some of the typical RM machines. RM can also be expressed as a branch of additive fabrication, which refers to the technologies employed to create physical models, prototypes, tools, or finished parts using 3-D scanning systems.

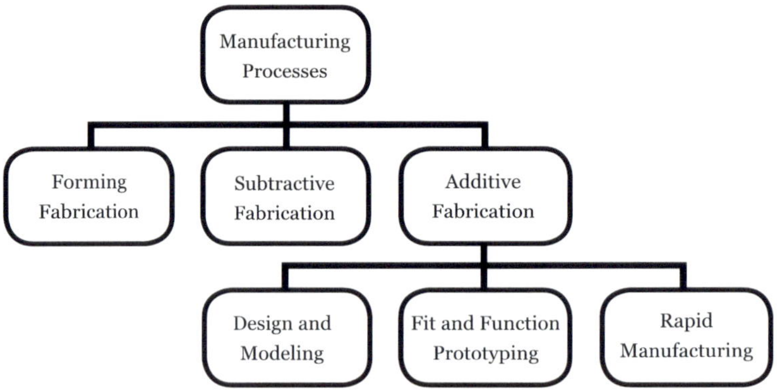


Figure: Manufacturing processes tree and rapid manufacturing (RM) position

## Computer-aided Rapid Manufacturing

Computer technology has served RM in many different aspects. In this section, tool path generation for RM by different CAD formats and computer-aided process selection are explained. The majority of the RP and RM processes use STL CAD format to extract the geometrical data of the model. The tool path generation from the STL file is explained in detail. Drawing exchange format CAD file for-mat is very complex and is mainly used either for data exchange between CAD software or tool path generation for the computer numerical control machining process. However, the tool path generation method that is explained in this section can be an alternative approach for RM processes. Information about the standard for the exchange of product model data, an increasingly popular CAD format, is provided in this section.

## Path Generation by use of Drawing Exchange Format File

- What is a DXF file : The DXF is a CAD data file format developed by Autodesk to transfer data between AutoCAD and other programs. Each of the eight sec-tions of a DXF file contains certain information about the drawing: HEADER section (for general information about the drawing), CLASSES section (for application-defined classes whose instances appear in the other sections), TABLES section (for definitions of named items), BLOCKS section (for describing the entities comprising each block in the drawing), ENTITIES section (for the drawing entities, including any block references), OBJECTS section (for the data that apply to nongraphical objects, used by AutoLISP and ObjectARX applications), and THUMBNAILIMAGE section.

- DXF for tool path generation: Because of the complexity of the DXF files, generating a tool path from these files for any automated machining or fabricating process is very difficult. In such a tool path generation mechanism, geometrical data need to be identified and extracted from many other data available in a DXF file that are not useful for a tool path. Data can be extracted from model that are designed as two-dimensional (2-D or wireframe) or two and half dimensional (2.5-D or surface) objects. Extracting data from DXF files containing 2.5-D objects is more complex than doing so from DXF files containing 2-D objects because for 2-D objects, the geometrical data are stored in the form of the object end points (e.g., start and end points of a line) while for 2.5-D objects, the geometrical data of a model (e.g., the surface of a sphere) are stored in the form of many small rectangles. In such case, for each rectangle x, y, and z coordination of all four corners are stored.

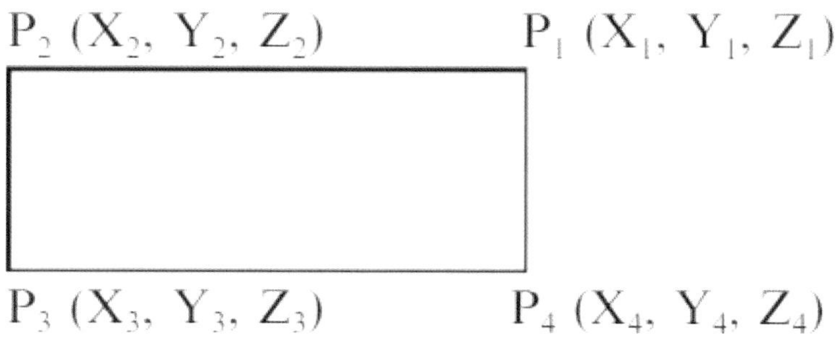

Figure: DXF file data storage format for a 2.5-D object

- DXF file generation: Illustration of the general steps to create a DXF file in the AutoCAD environment is given. As shown in this figure, after drawing the boundary of the object in wireframe format, a surface is generated to cover the entire outer boundary of the object. Then, after separating each face from the object, the entire data are stored in a DXF file. AutoCAD provides users with the flexibility to set a desirable number of meshes of the surface modeling in both horizontal and vertical directions.

- Generating tool path from DXF files: In automated controlling of the tool in any machine that uses the tool path, regardless of the process or the machine type, geometrical data need to be extracted from the CAD file. Then, the tool path is calculated and generated on the basis of the geometry of the model as well as process and tool specification. A tool in here can be a milling machine tool, a laser cutter head, a welding machine gun head, or an extruder nozzle. Figure shows the general steps of generating tool path from a DXF file.

As shown in, in addition to geometrical data, the desired quality of the final part affects the tool path output. Figure illustrates two different tool path configurations for the same CAD data.

- Tool radius compensation: If the tool radius is considered for the tool path generation, which is a must for almost all path-based processes, then the complexity of the tool path generation process becomes more complicated. In the tool radius compensation, the tool path is calculated for the center of the tool (not for the touching point of the tool and the part). Therefore, in the tool path generation process, both the curvature of the object at the tool and object touching point and the tool specification (size and shape) are affecting the center of the tool coordination. The curvature of the object at any point is shown by the normal vector. A normal vector is a vector (often unit) that is perpendicular to a surface.

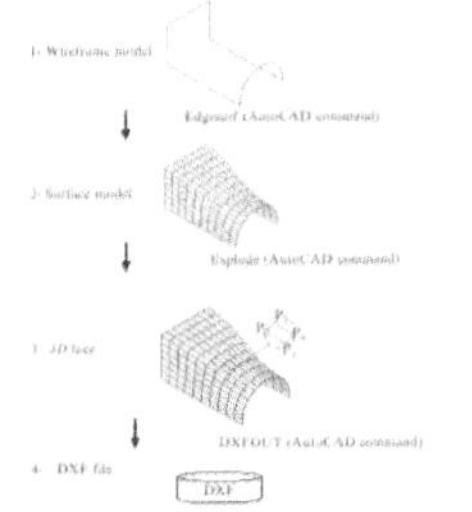

Figure: DXF file generation procedure for a 2.5-D object in Auto-CAD environment

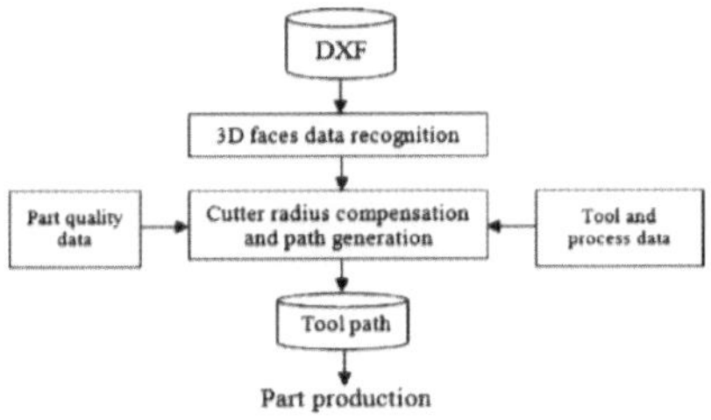


Figure: Tool path generation from DXF file

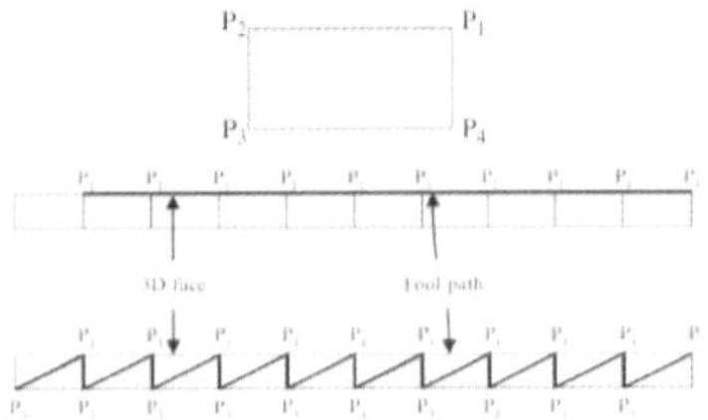

Figure: Two different tool path configurations for the same CAD data

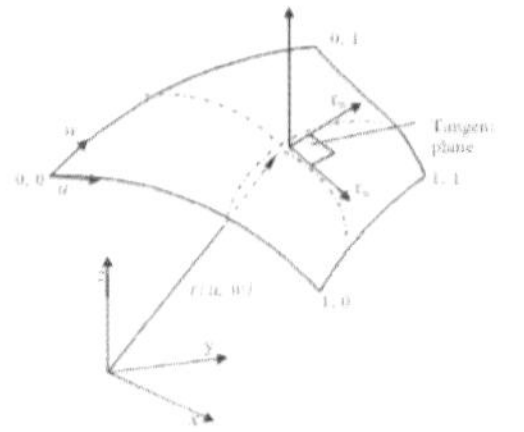

Figure: A surface and its normal vector

Specifications of the tools may cover a variety of shapes and sizes. Three of the most common geometries of the tools are shown. Most of the tools for machining and fabricating processes such as milling, extruding, welding, and laser beam are similar to one of these three geometries.

- Flat end tool: To calculate the position of the tool center for the tool path, the geometrical data of four corners of each rectangular 3-D face is used. In this process, the unit normal vector of the 3-D face is calculated, and then it is used to determine the tool center position). In the flat end tools, the tool center is located at the same height (z level) as the tool and object touching point. This fact simplifies the calculation for the z level of the tool position.

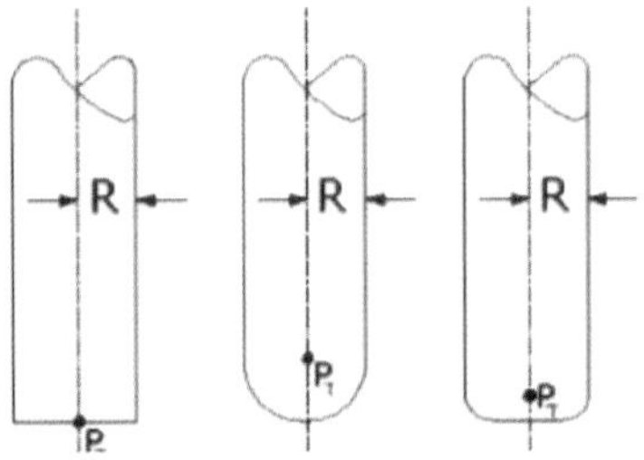


Figure: Three common geometries for tools: 1 – flat end, 2 – ball end, and 3 – tip radius

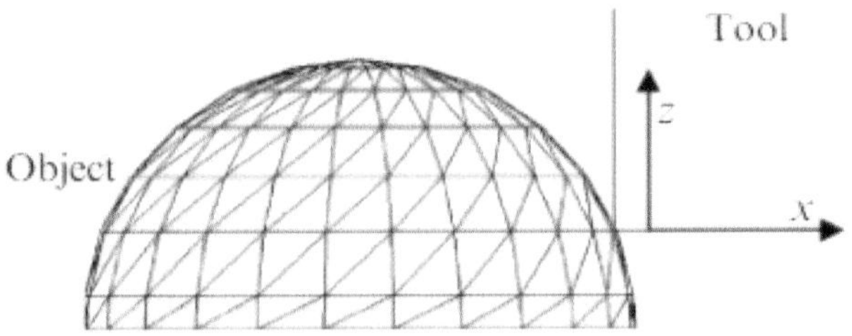


Figure: Flat end tool

$$\text{Normal vector} = r_u \times r_w = \begin{bmatrix} 1 & 1 & 1 \\ X_u & Y_u & Z_u \\ X_w & Y_w & Z_w \end{bmatrix}$$

$$= \left[ (Y_u Z_w - Z_u Y_w) \quad (Z_u X_w - X_u Y_w) \right]$$

$$= \left[ A \quad B \quad 0 \right]$$

$$r_u \cdot r_w = (A^2 + B^2)^{0.5}$$

$$\text{Unit normal vector} = N_1(U,W) = \left[ \frac{A}{(A^2+B^2)^{0.5}} \frac{B}{(A^2+B^2)^{0.5}} 0 \right]$$

- Tool Center:

$$X_a = X_1 + R\frac{A}{(A^2+B^2)^{0.5}}$$

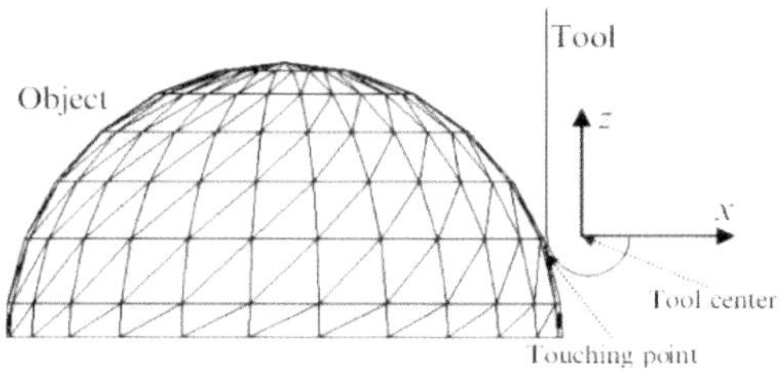


Fig. Ball end tool

$$Y_a = Y_1 + R\frac{A}{(A^2+B^2)^{0.5}}$$

$$Z_a = Z_1$$

- Ball end tool: The calculation of tool center position for ball end tools is very similar to that for flat end tools. The only difference is the z level of the tool center that, similar to x and y coordination, needs to be determined by the unit normal vector of the 3-D faces.

$$N_2(U,W) = \left[ \frac{A}{(A^2+B^2+C^2)^{0.5}} \frac{B}{(A^2+B^2+C^2)^{0.5}} \frac{C}{(A^2+B^2+C^2)^{0.5}} \right]$$
$$= N_2(n_x n_y n_z)$$

$$P_a = P_1 + R.N_2(n_x\, n_y n_z)$$

- Tool center:

$$X_a = X_1 + R\frac{A}{(A^2+B^2+C^2)^{0.5}}$$

$$Y_a = Y_1 + R\frac{B}{(A^2+B^2+C^2)^{0.5}}$$

$$Z_a = Z_1 + R\frac{C}{(A^2+B^2+C^2)^{0.5}}$$

- Tip radius tool : The geometry of this type of tool is the combination of the last two tools. The center part of this tool is a flat end tool while the edge of the tool has a fillet. Therefore, the position of the tool center is determined on the basis of the above two tools' calculations.

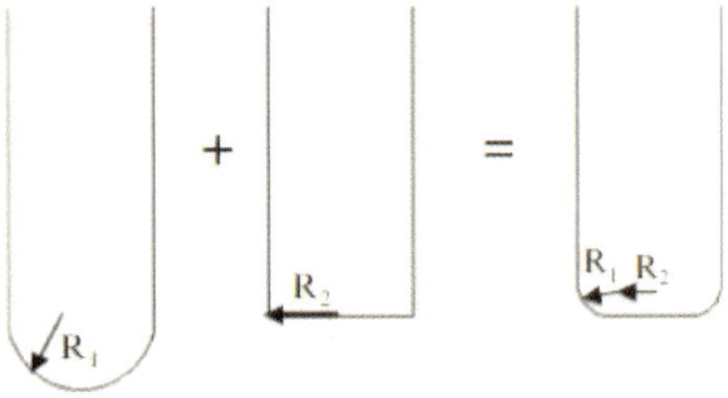


Figure: Geometry of the tip radius tool

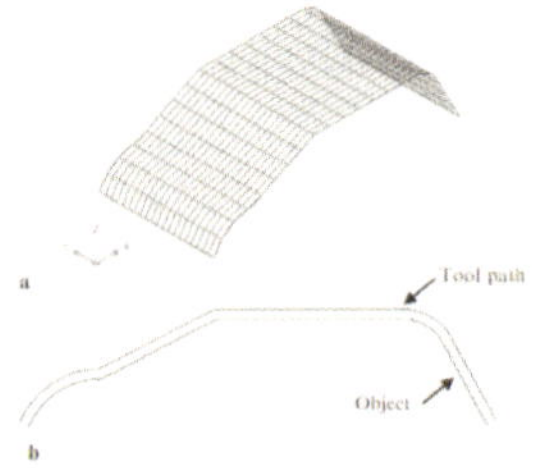


Figure: A meshed CAD object and b the position of the object and tool path

$$P_a = P_1 + R_1 N_1 + R_2 N_2$$

- Tool Center:

$$X_a = X_1 + R_1 \frac{A}{\left(A^2 + B^2 + C^2\right)^{0.5}} + R_2 \frac{A}{\left(A^2 + B^2\right)^{0.5}}$$

$$Y_a = Y_1 + R_1 \frac{A}{\left(A^2 + B^2 + C^2\right)^{0.5}} R_2 \frac{B}{\left(A^2 + B^2\right)^{0.5}}$$

$$Z_a = Z_1 + R_1 \frac{C}{\left(A^2 + B^2 + C^2\right)^{0.5}}$$

Figure illustrates a meshed CAD object (a) and the position of the object and tool path. At the end, it is necessary to mention that because of the complexity and limitations of the DXF files, it is very uncommon to use DXF to produce a tool path for RP, RM, or even 3-D and 2.5-D CNC machine code. Its application in CAM is usually limited to 2-D applications such as drilling.

## Path Generation by use of STL File

Every RM and RP system has its own specifications. The part boundary form, part filling method, and part separation from the surrounding material determine the tool path pattern for every layer. This tool path pattern could be a robotic movement in XY plane for FDM or contour crafting machines, a laser pattern for material solidification in SLA and selective laser sintering (SLS) machines, or a laser cutter pattern for a Laminated Object Manufacturing (LOM) machine. These processes require different tool path pattern generation strategies.

Therefore, unlike CNC standard tool path files (e.g. APT and G-Code), there is no standard tool path file for RP systems. Therefore, most of the new RM and RP require a new tool path generator or modification to the previous systems. In this section, a tool path generation for the selective inhibition of sintering (SIS) process is presented. The software that is developed for this can be modified and adjusted for many other RP and RM processes. This system uses STL files with the ASCII format as input and works in two steps.

## Step 1—Slicing Algorithm

In Step 1, the STL file is read as input. Slice files are then generated by executing the slicing algorithm. Only the intersection of those facets that intersect current $Z = z$ is calculated and saved. In this step, one facet is read at a time from the STL file. Then the intersection lines of this facet with all XY planes for $Z_{min} \leq z \leq \text{Min}\{Z_{min}\ Max\{z_A, z_B, z_c\}\}$ are calculated. The intersection lines are stored in the specified file for the associated z increment.

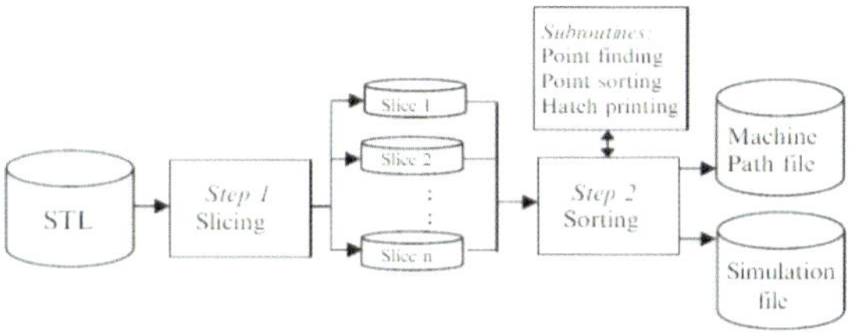


Figure: The two steps of slicing and tool path generation

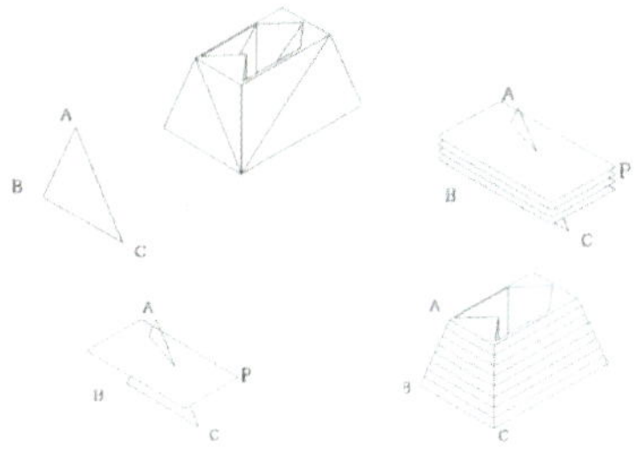

Figure: Slicing algorithm steps

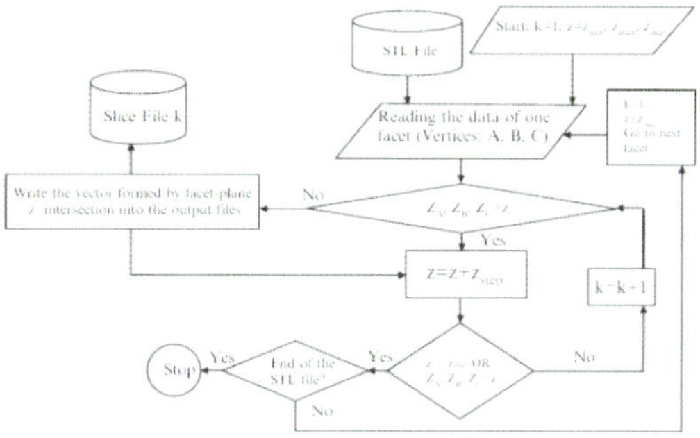


Figure: The slicing algorithm

This results in one intersection line on each XY plane. By repeating this process for all facets, a set of slices is generated. This algorithm saves the data of only one facet in the computer memory; therefore only a small amount of computer memory is needed, and there is no practical limitation on the model size. In this step, each slice is saved in a separate file on the disk. This guarantees that Step 2 is run much faster than when all slices are saved in a single file.

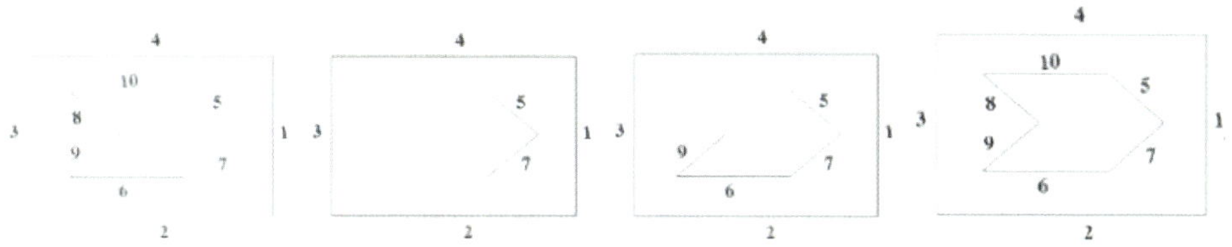

Figure: Tool path generation steps

## Step 2—Tool Path Generation

After the completion of the slicing process, a set of vectors becomes available in each z increment. These vectors are not connected and are not in sequence. In the tool path generation process, the software starts from one vector and tries to find the next connected vector to this vector. Then it does the same for the newly found vector until it reaches the start point of the first vector (in the closed loop cases) or finds a vector with no leading attachment (in faulty STL files containing disconnections). To sort the vectors, the algorithm reads one vector at a time from a slice file and writes it to another file. This file is either a path file, when one vector is connected to the previous vector, or a temp file, when the vector is not connected to the previous vector. Therefore, the sorting process does not need a large amount of memory to sort the data, and there is no limitation on the number of vectors in a slice and on input file size. In addition, unlike many other slicing algorithms that cannot handle disconnections caused by faulty facets, this algorithm can generate a tool path even with disconnection errors in the STL file. At disconnection instances the system sends a message to a log file and turns the printer off and starts from a new vector. In either case, the printer is turned off and the system starts printing from another start point. Also for each selected vector, the possibility of hatch intersection points is investigated.

At the end of the path generation process for one slice, the hatch intersection points are sorted and written into a tool path file. After the arrangement of all vectors in one slice (z increment), the process starts arranging the vectors of the next slice. This process continues until all vectors in all slices are sorted.

## Implementation

Slicing and tool path generation algorithms have been implemented in the C programming language. The software has been successfully tested for several medium and large STL files up to 200 MB on different PCs and laptops.

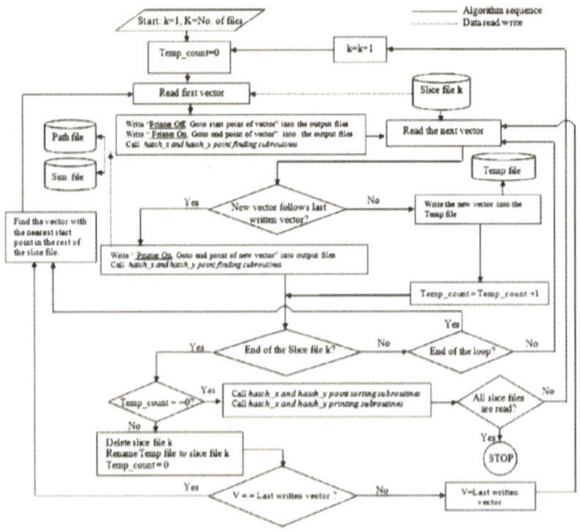


Figure: Tool path generation algorithm

## Path Generation by use of STEP File

STEP, introduced by PDES, Inc., is a neural format for exchanging product data among all the people or organizations who contribute to marketing, design, manufacturing, and other activities in the product life cycle. STEP makes an independent platform to access, share, and manipulate the product information of its life cycle.

There are other types of product data exchange (PDE) models like IGES, DXF, DWG, VHDL, and STL which are used by many industries all over the world but STEP is known as a superior standard and is becoming highly popular over the other formats such as IGES.

## Application Protocols

STEP is developed by a specific language of EXPRESS and includes a number of manageable and functional sections referred to as application protocol (AP).

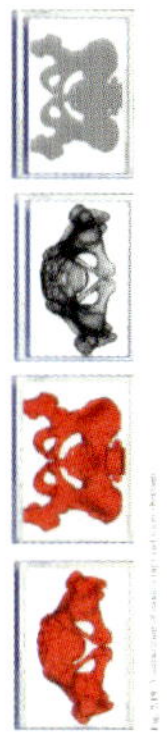

Figure: Visualization of models (up) and slices (bottom)

Figure: Visualization of CAD file and the tool path in the Auto-CAD environmen

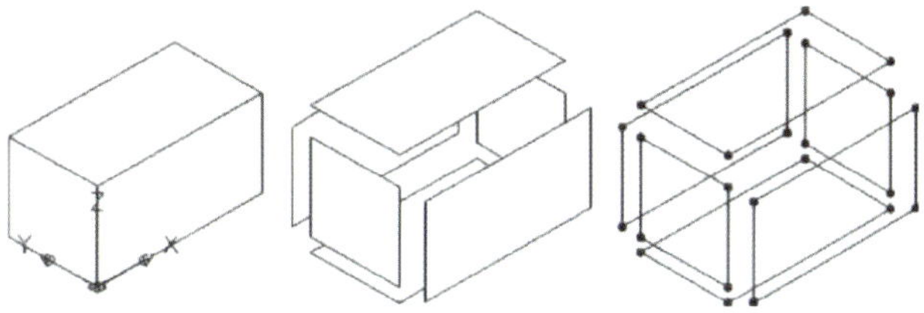

Figure: B-rep model of a simple cube: faces, edges, and vertices depicted in the model

AP is in charge of an application of STEP. For example, AP224 or ISO 10303–224 are mechanical product definitions for process planning using machining features; AP203 is configuration-controlled 3-D designs of mechanical parts and assemblies. For the time being, only 22 APs have been approved as International Standards; however, it is expected that there will be hundreds of APs in the future.

STEP file contains all the required information for design, process planning, and other downstream activities; the implementation of STEP tools depends on the requirements and problem area in the organization. For process planning, a mechanical part based on its machining features, AP224, will be applicable as it classifies the features and can store and manage the data needed for each feature.

## Boundary Representation Model

Some APs contain boundary representation (B-rep) models for mechanical parts. For every 3-D part, B-rep expresses the surface boundary, which consists of the part's geometrical and topological information. Vertices, edges, and faces information is also represented by a B-rep model.

## Using STEP for Tool Path Generation

No direct usage of STEP file to generate the tool path in RM has been reported. But as indicated, some of the application protocol contains B-rep information for the products. This information model can be extracted and used for tool path generation. Consider a part meshed with triangles and sliced in STL file format is replaced with a part having information of all the edges, faces, and vertices. Laser trajectory will be obtained by introducing the points of outer surfaces. This may not be applicable only for free-form parts with complex geometry as their B-rep information is not complete on the surfaces.

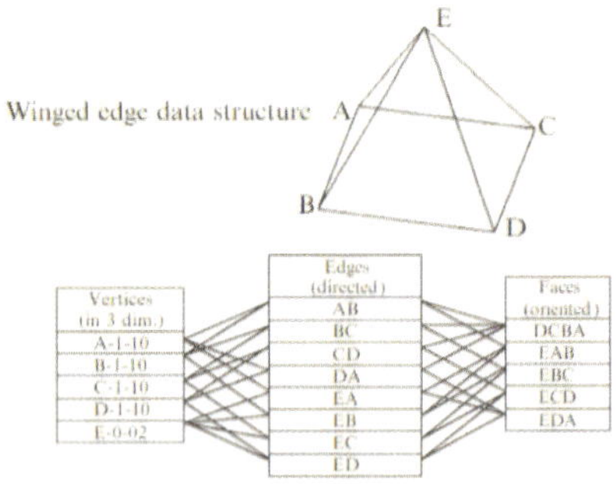


Figure: Boundary representation (B-rep) of a simple part

Figure: A typical part to be rapid-manufactured

Even with a B-rep model, slicing is required to determine the exact intersection points and feed the control system of laser scanning. AP240, numerical control process plans for machined parts (ISO 10303–240:2005), is a part of ISO 10303 that specifies information requirements for the exchange, archival, and sharing of computer-interpretable numerical control (NC) process plan information and associated product definition data but it does not include specific machine tool controller codes. AP204, AP207, AP210, AP214, AP224, AP227, and AP240 are the applications which have boundary information of the corresponding parts and may be used for tool path generation.

```
ISO-10303-21;
HEADER;
FILE_DESCRIPTION(('This is a test file'),'1');
FILE_NAME('test','2007-8-18',('Authors'),('Authors'),'NA','OK');
FILE_SCHEMA('ARM224');
ENDSEC;
DATA;
#1 = CARTESIAN_POINT((0.176777,0.176777,0.5));
#2 = VERTEX_POINT(#1);
#3 = CARTESIAN_POINT((0.,0.,0.5));
#4 = DIRECTION((0.,0.,–1.));
#5 = DIRECTION((0.176777,0.176777,0.));
#6 = AXIS2_PLACEMENT_3D(#3,#4,#5);
#7 = CIRCLE(#6,0.25);
#8 = EDGE_CURVE(#2,#2,#7,.T.);
#9 = ORIENTED_EDGE(*,*,#8,.F.);
#10 = EDGE_LOOP((#9));
#11 = FACE_BOUND(#10,.T.);
#12 = CARTESIAN_POINT((0.176777,0.176777,–0.5));
#13 = VERTEX_POINT(#12);
#14 = CARTESIAN_POINT((0.,0.,–0.5));
#15 = DIRECTION((0.,0.,1.));
```

Figure: A piece of a STEP file

## Rapid Manufacturing Process Selection and Simulation

Because of the inherent strengths and weaknesses of all RM processes, choosing the method best fitted to economical and technical objectives is a serious problem. This process selection is a multi-criteria decision making. Along with satisfying customer requirements, the functional product is expected to be manufactured to an adequate quality and at most reasonable cost in the shortest possible time. Other criteria such as product recyclability or serviceability may be considered.

One of the approaches for making such simulation in RP and RM uses computer software in order to create a virtual prototype. It is due to high material cost and prototyping of the real prototypes. Simulators help engineers to evaluate the performance of the process and minimize the number of repetitions to reach an appropriate prototype. One successful attempt to introduce such simulators was made by Choi and Samavedam. They developed a simulation software program to show the RM process layer-by-layer with SLS process. It shows also the relation between process parameters and time and accuracy of the product. The main advantages of running simulation in RP and RM are listed below:

- The cost of materials especially for some processes such as SLS is notable.
- The process of making the product is usually time-consuming.
- The energy consumption and equipment depreciation are high.
- The quantitative parameters of the prototype will be easily extracted.
- The product information may be shared with other persons or research centers.

VIRAPS (Virtual Rapid Prototyping System) is a simulation software program developed by Visual Basic and simulates some of the most common processes of RP and RM, like FDM, SLS, SLA, and LOM. Here we describe how VIRAPS works and the outcomes.

There are many criteria to consider in choosing the right RM process. Since it is a multi-criteria decision, an analytic hierarchy process (AHP) methodology could be applied to rank the most appropriate process according to the customer's criteria. Using software, RM processes are compared to each other on the basis of some common characteristics such as average time, cost, and quality of the finished products. The priority of these criteria is also determined by customer and the highest ranked RM process and machine is selected accordingly.

As a case study, SLS is considered for simulation. Suppose that a conic-shaped product is manufactured by SLS. The important inputs for this process are as follows:

- Part information (maximum dimensions, average slope, and layer thickness),
- Laser specification (laser type, laser power, beam diameter, and scan speed),
- Powder selection (here is steel-bronze),
- Setup time for each layer (the time required between creating each layer for setup),
- Cost parameters (direct cost, operation cost, and material cost.

The software also indicates the maximum and minimum allowed for some parameters such as scan speed or beam diameter and product weight according to the machine capabilities. Moreover, some parameters, such as type of laser, are automatically selected by the software.

The outputs are shown in the figure. This product will be made in 6.5 h and with the accuracy ratio of 73 and efficiency of 47%. The total estimated cost is $344 and the number of layers according to the best orientation of the part is 500. These results are calculated on the basis of input information. For instance, the time in SLS is formulated below:

$$\text{Velocity}\,(\text{v}) = \frac{P_1(1-R)}{pd_b l_m \left[C_p(T_m - T_b)\right]}$$

where velocity $(v)$ = velocity of laser (mm/s), $P_1$ = power of laser emission (watt), $R$ = reflection ratio of laser reflector mirror, $\rho$ = material density (g/mm 3 ), $d_b$ = beam diameter (mm), $l_m$ = layer thickness (mm), $C_p$ = specific heat capacity $(\text{J}/\text{k})$, $T_m$ = material melting point (K), and $T_b$ = laser scanning time (s).

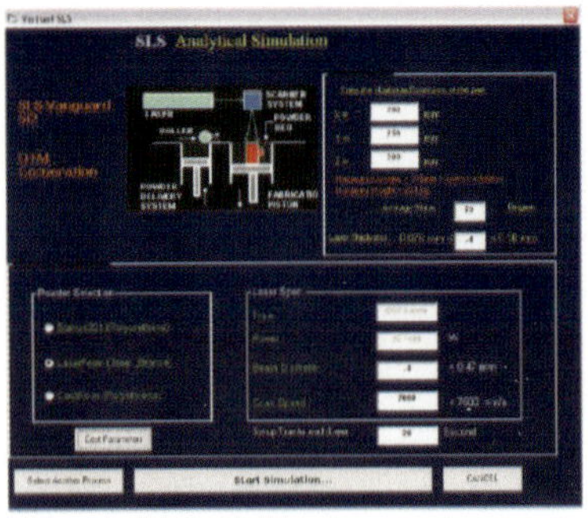


Figure: The SLS analytical simulation interface

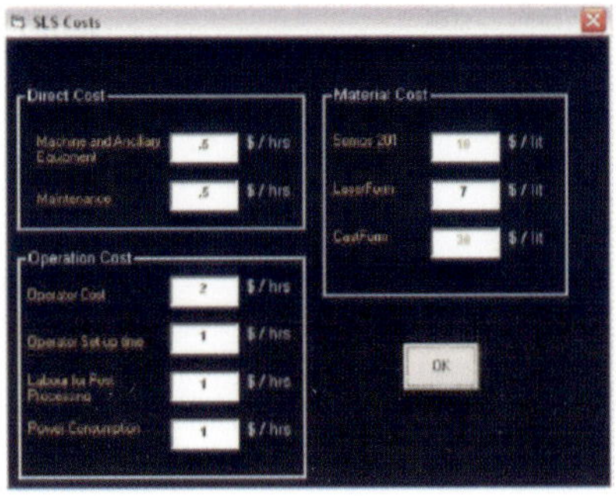


Figure: Software interface for process cost items

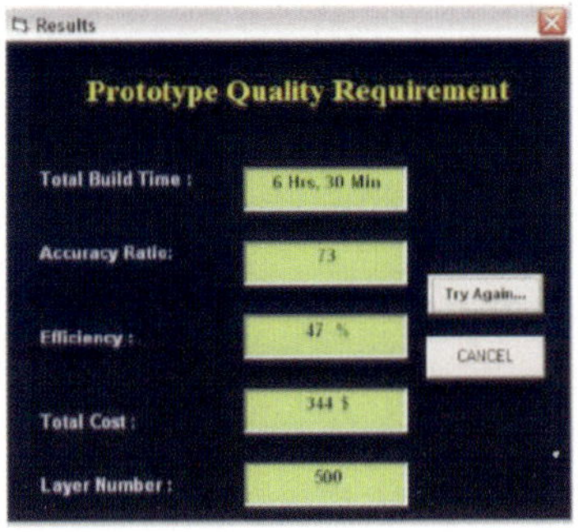


Figure: Process simulation outcomes

On the one hand, the time required for building the whole part is the time taken for making all the layers and the time of setup for each layer. On the other hand, the time of scanning each layer $T_1$ is $\frac{L_d}{L_v}$ in which $L_d$ is the laser scanning distance and $L_v$ is the laser scanning speed.

$$\text{Build time(total)} = \sum_{\neq 1}^{N} T_{1i} + T_s N_1$$

where build time (total) = total time for building layers, $T_{1i}$ = scanning time for layer $i$, $T_s$ = setup time for each layer, and $N_1$ = number of layers.

Setup time is separately obtained by the following relations:

$$\text{Setuptime(Ts)} = \text{T}_{wd} + T_d + T_{wr} + T_h$$

where $T_{wd}$ = time required for moving the part bed downward (s), $T_d$ = time required for pouring a layer of material (s), $T_{wr}$ = time required for moving the part bed upward (s), and $T_h$ = time required for preheating the material (s).

Therefore, assuming each layer has thickness of $l$ and the distance scanning of laser for layer $i$ is $d_{si}$, for a part with height of h, we have

$$\text{Build time(total)} = \left(\frac{h}{I_m}\right)T_s + \frac{\sum_{i=1}^{N_1} d_{si}\left(\frac{I}{I_m}\right)}{L_v}$$

## Applications of Rapid Manufacturing

RM is widely used for both large- and small-scale products and components for a variety of applications in many different fields. The main applications of the RM can be categorized as follows.

## Tooling and Industrial Applications

Fabrication of metal casting and injection mold has been one of the main applications of RM in recent years, and has been addressed in the literature.

## Aerospace

RM products have found efficient applications in spacecraft structures (mirrors, structural panels, optical benches, etc.). They are made up of different titanium and aluminum alloys (with granulated powder) and other materials such as silicon carbide, metal ceramic composites (SiC/Alu, ferrous materials with SiC), and carbon fiber reinforced polymer.

## Architecture and Construction

In today's construction, CAD/CAM technology, industrial robots, and machines which use direct numerical control and RM open up the architectural possibilities. The Buswell's research into the use of mega-scale RM for construction, Pegna's investigation of solid freeform construction, and Khoshnevis's development of a new layer-by-layer fabrication technique called contour crafting are some of the attempts to apply RM in architecture and construction industry.

## Military

The production costs of military complex airframe structures were notably reduced where direct metal deposition, an RM technique, was applied to build limited number of metallic parts.

## Medical Applications

Medical application of RM is mainly due to its capability to build uniquely shaped products having complex geometry. Medical products like implants, dentures, and skull and facial bones are the components that often vary from one person to another. Some examples are (1) custom-made orthodontic appliances production using proprietary thermoplastic ink jetting technology and (2) ear prostheses and burn masks for patients by using thermo-jet printing and FDM. Other applications have been reported to manufacture patient-specific models (lead masks) as well as protective shields in cancer treatment by using 3-D photography and metal spraying technology. Forming a prosthetic glove of nontoxic materials to cover a patient's damaged hand is another medical application.

## Electronics and Photonics

A new RM methodology based on a direct-write technique using a scanning laser system to pattern a single layered SU-8 for fabrication of embedded microchannels has been reported by Yu. Nijmeijer explain how microlaminates, which are widely used in ceramic capacitors of electronic devices, are being made by an RM process: centrifugal injection casting.

## Rapid Manufacturing's Advantages and Disadvantages

RM conducted in parallel batch production has a large advantage in speed, cost, and quality over alternative manufacturing techniques such as laser ablation or die casting. RM changes the cost models in conventional supply chains and has a key role in producing and supplying cost-effective customized products. Consequently, RM's popularity is growing on a daily basis. According to a Wohlers Associates survey, RM's applications in additive processes grew from 3.9% in 2003 to 6.6% in 2004 and 8.2% in 2005.

### Advantages

RM's advantages can be studied from several design-related perspectives.

- Design complexity: One major benefit of the additive manufacturing processes is that it is possible to make parts of virtually any geometrical complexity at no extra cost while in every conventional manufacturing technique there is a direct link between the complexity of a design and its production cost. Therefore, for a given volume of component, it is possible to get the geometry (or complexity) for "free," as the costs incurred for any given additive manufacturing technique are usually determined by the time needed to build a certain volume of part, which, in turn, is determined by the orientation that the component is built in.
- Design freedom: The advent of RM will have profound implications for the way in which designers work. Generally, designers have been taught to design objects that can be made easily with current technologies—this being mainly due to the geometry limitations of the available manufacturing processes. For molded parts, draft angles, constant wall thickness, location of split line, etc. have to be factored into the design. Because of the advancements in RM, geometry will no longer be a limiting factor in design.
- New design paradigm: RM has simplified the interaction between mechanical and aesthetic issues. With current RM capabilities, industrial designers can design and fabricate the parts without the need to consider issues such as draft angle and constant wall thickness that are needed for processes such as injection molding. Similarly, mechanical designers are able to manufacture any complexity of product they require with minimum education in the aesthetic design field.

### Disadvantages

Like any other immature technology, RM has drawbacks and limitations, which preclude its widespread commercial application. Some main disadvantages are as follows:

- Material cost: Today, the cost of most materials for additive systems is about 100–200 times greater than that of those used for injection molding.
- Material properties: Thermoplastics from LS have performed the best for RM applications. However, a limited choice of materials is available. Actually, materials for

additive processes have not been fully characterized. Also, the properties (e.g., tensile property, tensile strength, yield strength, and fatigue) of the parts produced by the RP processes are not currently competitive with those of the parts produced by conventional manufacturing processes. Ogando reports that the properties of a part produced by the FDM process with acrylonitrile butadiene styrene (ABS) material are about 70–80% of a molded part. In contrast, in some cases, results are more promising for the RP processes. For example, the properties of the metallic parts produced by the direct metal laser sintering (DMLS) process are "very similar to wrought properties and better than casting in many cases." Also, in terms of surface quality, even the best RM processes need secondary machining and polishing to reach acceptable tolerance and surface finish.

- Support material removal: When production volumes are small, the removal of support material is usually not a big issue. When the volumes are much higher, it becomes an important consideration. Support material that is physically attached is of most concern.
- Process cost: At present, conventional manufacturing processes are much faster than additive processes such as RM. Based on a comparison study by the Loughborough University, the SLA of a plastic part for a lawn mower will become economical at the production rate of 5,500 parts. For FDM, the breakeven point is about 6,500 parts. Nevertheless, injection molding is found to be more economical when larger quantities must be produced.

## Rapid Manufacturing Errors

One of the main challenges in RP and RM is deviation from the CAD part model. The accuracy will be obviously enhanced by increasing the number of layers (decreasing the layer thickness) but the manufactured part is never identical to its CAD file because of the essence of layer-by-layer fabrication. Three major RM errors are as follows.

## Preprocess Error

While converting a CAD to standard tessellation language (STL) format (as machine input), the outer surface of the part is estimated to some triangles and this estimation causes this type of error, especially around the points with higher curvature (lower radius). Meshing with smaller triangles may diminish this error. However, it requires more time to process the file and a more complicated trajectory for laser.

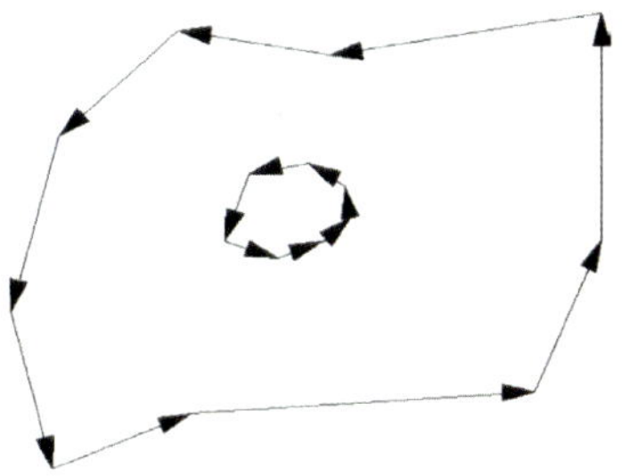

Figure: More triangles in more edges in each layer.

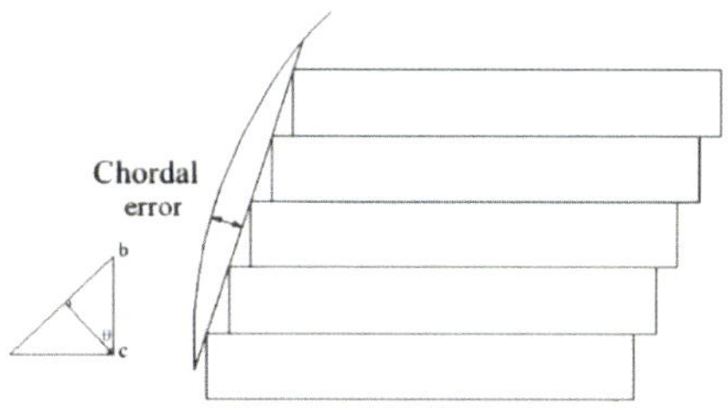


Figure: This error can be estimated as bc cos θ.

## Process Error

As shown in Figure, slicing the part causes a new type of error (chordal error) while building the part layer-by-layer. This error depends on the layer thickness and the average slope angel of the part. To minimize bc cos $\theta$, the thickness of layer, bc, in the curved area should be decreased and it leads to lower step-stair effect and a more accurate product. Poor laser scanning mechanism may prompt another error during the process. The accuracy of laser beam emission and its angle with the part surface affect product quality so to make it smaller, the equipments and tools should be carefully inspected.

## Postprocess Error

Two other types of errors that arise after the process are shrinkage and warpage errors. The product usually shrinks while cooling and this makes a deviation from its original design. Overestimation of CAD file at the design stage regarding the material properties and heating factors can reduce such an error. Warpage error is due to disparate distribution of heat in the part. Thermodynamic and binding force models are usually required to estimate and obviate this error.

## References


- Manufacturing-process-meaning-and-types: engineeringarticles.org, Retrieved 09 July 2018
- What-is-near-net-shape: wisegeek.com, Retrieved 24 April 2018
- Metal-casting-definition-types-and-processes: engineeringarticles.org, Retrieved 13 June 2018
- Additive-manufacturing-vs-subtractive-manufacturing: creativemechanisms.com, Retrieved 14 May 2018
- Make-parts-fast-handbook-subtractive-prototyping-2011: makepartsfast.com, Retrieved 29 July 2018
- Rapid-Manufacturing-226514527: researchgate.net, Retrieved 19 March 2018

# 3

# Casting Processes

Casting is a manufacturing process characterized by the design of complex shapes by pouring a liquid material into a hollow cavity, which is then allowed to solidify. The topics elaborated in this chapter on evaporative-pattern casting, permanent mold casting, die casting, continuous casting, centrifugal casting, etc. will help in providing a better perspective about casting processes and techniques.

## Casting

Casting is defined as a manufacturing process in which molten metal pure into a mold or a cavity of desire shape and allow to solidify which form a predefine shape. This process is widely used to manufacture complex parts which cannot be made by other processes. All major parts like bed of lathe machine, milling machine bed, IC engine component etc. are made by this process.

### Casting Terminology

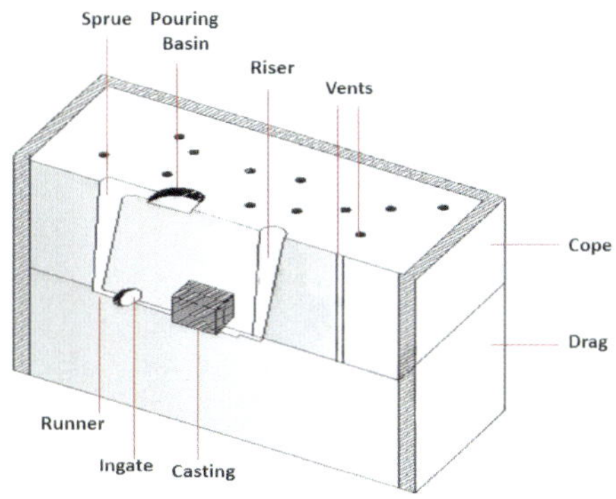


**Casting Terminology**

The moulding flask is used to hold the sand mould. The sand mould has desire cavity of object which to be casted. The sand is rammed into the flask to create sand mould in which metal is poured to get desire shape. It is created into minimum two pieces which allows removal of pattern easily.

The upper part of moulding flask is known as cope.

The lower part of moulding flask is known as drag.

When the moulding flask made into more than two parts, the intermediate parts are known as cheeks. These are used in complex casting.

Pattern is replica of object to be created. It is made by either wood, wax or other suitable material. It is placed into moulding flask and sand rammed over it which forms an object cavity into sand.

It is a funnel shape cavity at the top of the mould. The metal is poured into pouring basin from where it is supplied at different parts of mould.

Runner is a horizontal passage of molten metal. It connects sprue to getting system. Normally it is situated at lower half of mould.

Riser is used as reservoir of molten metal when pouring of molten metal has stopped. When the cavity is filled by molten metal, the pouring is stopped which allows solidifying object. During solidification, volumetric shrinkage takes place which reduces the desire size and shape of object. The riser is provided into the mould which supplies the molten metal to remove effect of volumetric shrinkage during solidification. These are further divided into top riser, blind riser, side riser etc.

It is a passage which connects pouring basin to the runner. It controls the flow of molten metal from pouring basin. It is tapered in shape.

It is the entry point through which molten metal enters into the actual mould cavity.

Core is used to cast hollow cavity. It is also a sand structure and placed at right place into mould cavity where hollow part is to be created. The metal poured into mould cavity does not fill the part at which core is placed thus form a hollow cavity.

These are supporting components of core. These used to support and hold the core into mould cavity. These protect the core from various forces encounter in casting.

Chills are generally solid metal pieces which are placed into cavity to increase cooling rate. Mainly it is used to create direction solidification of molten metal. They have high thermal conductivity.

These are small passages made in mould which allow to escape the gases during solidification.

## Casting Processes

The major casting processes used in the manufacturing industry are the following:

### Sand Casting

Sand casting is normally used for the production of large parts, by filling a molten metal into the mold cavity that has been shaped from natural or synthetic sand. The cavity is created by the utilization of a pattern, generally made of wood or metal that is of the same shape and dimensions as the actual part. The pattern is prepared slightly oversize due to which the cavity is also a little larger and compensates for the contraction of molten metal during cooling. Surface of the sand castings is normally rough with surface impurities for which a machining allowance is included.

## Die Casting

In this process metal is forced into the mold at a high pressure that ensures production of identical parts, a better surface finish, and an increased dimensional accuracy. Some parts produced by die casting even do not require machining after casting, or may require only a light machining to achieve the desired dimensions. Defects of porosity are found more often in large castings because of entrapped air and the solidification of melt before it reaches the boundaries of the cavity. Parts with a uniform wall thickness can be more accurately produced by die casting. Die casting molds are expensive since these are made from hardened steel and because a longer time duration is required for their production.

## Centrifugal Casting

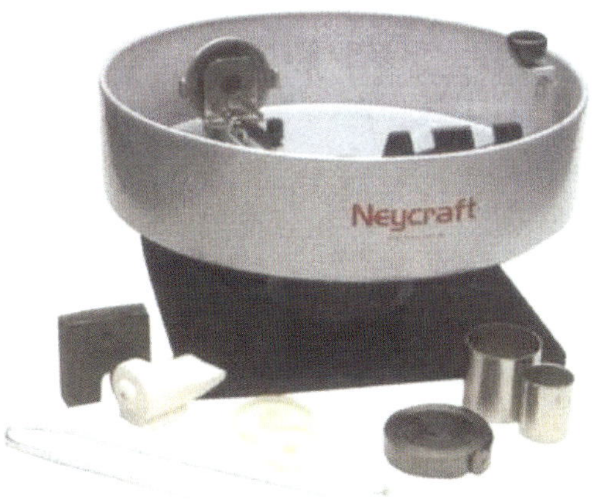


This is a casting technique that has an extensive range of industrial applications including the casting of machine fittings where durability of the finished product is important. Television picture tubes, spherical glass objects, pipes, flywheels, and boilers are also produced by centrifugal casting. As the molten metal is poured, a perman ent mold spins at high speeds around its axis. The molten metal moves towards the mold walls due to centrifugal force, solidifies after cooling, producing a fine part. The external diameter of the casting has fine grains, and is resistant to atmospheric corrosion.

## Investment Casting

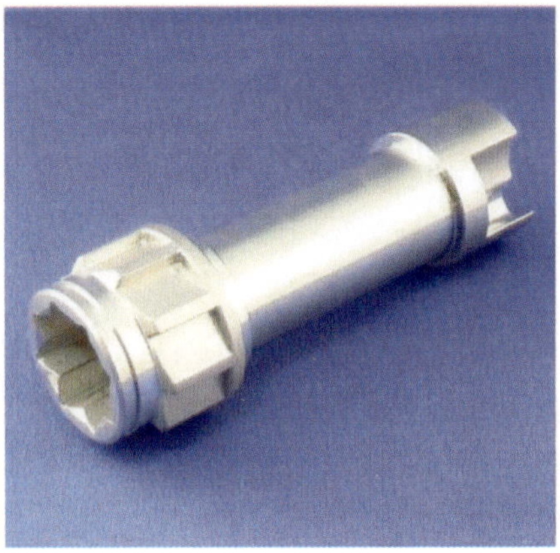

Investment casting is an ancient manufacturing process used for metals that are difficult to be machined or fabricated. It is also used for the manufacture of parts that cannot be formed by usual manufacturing techniques like turbine blades or components of airplane that are subjected to high temperatures. This process provides an excellent dimensional accuracy and surface finish. The pattern is made of wax or other substance that is melted, leaving behind a cavity which is filled with the material of the part being produced.

Casting has following advantages over other manufacturing process.

- It can create any complex structure economically.
- The size of object doesn't matter for casting.
- The casting objects have high compressive strength.
- All structure made by casting has wide range of properties.
- This can create an accurate object.
- All material can be cast.
- It creates isotropic structure.
- It is cheapest among all manufacturing processes.
- Composite component can be easily made by casting.

Along these advantages, casting has following disadvantages.

- It gives poor surface finish and mostly requires surface finish operation.

- Casting defects involves in this process.
- It gives low fatigue strength compare to forging.
- It is not economical for mass production.

## Casting (Metalworking)

In metalworking, casting involves pouring liquid metal into a mold, which contains a hollow cavity of the desired shape, and then allowing it to cool and solidify. The solidified part is also known as a casting, which is ejected or broken out of the mold to complete the process. Casting is most often used for making complex shapes that would be difficult or uneconomical to make by other methods.

Molding or moulding (see spelling differences) is the process of manufacturing by shaping liquid or pliable raw material using a rigid frame called a mold or matrix. This itself may have been made using a pattern or model of the final object.

A mold or mould is a hollowed-out block that is filled with a liquid or pliable material like plastic, glass, metal, or ceramic raw materials. The liquid hardens or sets inside the mold, adopting its shape. A mold is the counterpart to a cast. The very common bi-valve molding process uses two molds, one for each half of the object. Piece-molding uses a number of different molds, each creating a section of a complicated object. This is generally only used for larger and more valuable objects.

The manufacturer who makes the molds is called the moldmaker. A release agent is typically used to make removal of the hardened/set substance from the mold easier. Typical uses for molded plastics include molded furniture, molded household goods, molded cases, and structural materials.

Casting processes have been known for thousands of years, and widely used for sculpture, especially in bronze, jewellery in precious metals, and weapons and tools. Traditional techniques include lost-wax casting, plaster mold casting and sand casting.

Metal casting is one of the most common casting processes. Metal patterns are more expensive but are more dimensionally stable and durable. Metallic patterns are used where repetitive production of castings is required in large quantities.

Casting is a 6000 year old process. The oldest surviving casting is a copper frog from 3200 BC.

The modern casting process is subdivided into two main categories: expendable and nonexpendable casting. It is further broken down by the mold material, such as sand or metal, and pouring method, such as gravity, vacuum, or low pressure.

Casting is a manufacturing process by which a liquid material is usually poured into a mold, which contains a hollow cavity of the desired shape, and then allowed to solidify. The solidified part is also known as a casting, which is ejected or broken out of the mold to complete the process. Casting materials are usually metals or various cold setting materials that cure after mixing two or more components together; examples are epoxy, concrete, plaster and clay.

Casting is most often used for making complex shapes that would be otherwise difficult or uneconomical to make by other methods.

Epoxy is the cured end product of epoxy resins, as well as a colloquial name for the epoxide functional group. Epoxy resins, also known as polyepoxides are a class of reactive prepolymers and polymers which contain epoxide groups. Epoxy resins may be reacted (cross-linked) either with themselves through catalytic homopolymerisation, or with a wide range of co-reactants including polyfunctional amines, acids (and acid anhydrides), phenols, alcohols, and thiols. These co-reactants are often referred to as hardeners or curatives, and the cross-linking reaction is commonly referred to as curing. Reaction of polyepoxides with themselves or with polyfunctional hardeners forms a thermosetting polymer, often with strong mechanical properties as well as high temperature and chemical resistance. Epoxy has a wide range of applications, including metal coatings, use in electronics / electrical components, high tension electrical insulators, fiber-reinforced plastic materials, and structural adhesives. Epoxy resin is employed to bind gutta percha in some root canal procedures.

Epoxy resins are low molecular weight pre-polymers or higher molecular weight polymers which normally contain at least two epoxide groups. The epoxide group is also sometimes referred to as a glycidyl or oxirane group.

A wide range of epoxy resins are produced industrially. The raw materials for epoxy resin production are today largely petroleum derived, although some plant derived sources are now becoming commercially available (e.g. plant derived glycerol used to make epichlorohydrin).

Epoxy resins are polymeric or semi-polymeric materials, and as such rarely exist as pure substances, since variable chain length results from the polymerisation reaction used to produce them. High purity grades can be produced for certain applications, e.g. using a distillation purification process. One downside of high purity liquid grades is their tendency to form crystalline solids due to their highly regular structure, which require melting to enable processing.

An important criterion for epoxy resins is the epoxide content. This is commonly expressed as the epoxide number, which is the number of epoxide equivalents in 1 kg of resin (Eq./kg), or as the equivalent weight, which is the weight in grams of resin containing 1 mole equivalent of epoxide (g/mol). One measure may be simply converted to another:

Equivalent weight (g/mol) = 1000 / epoxide number (Eq./kg)

## Plaster, Concrete, or Plastic Resin

Plaster and other chemical setting materials such as concrete and plastic resin may be cast using single-use waste molds, multiple-use 'piece' molds, or molds made of small rigid pieces or of flexible material such as latex rubber (which is in turn supported by an exterior mold). When casting plaster or concrete, the finished product is, unlike marble, unattractive, lacking in transparency, and so it is usually painted, often in ways that give the appearance of metal or stone. Alternatively, the first layers cast may contain colored sand so as to give an appearance of stone. By casting concrete, rather than plaster, it is possible to create sculptures, fountains, or seating for outdoor use. A simulation of high-quality marble may be made using certain chemically-set plastic resins (for example epoxy or polyester) with powdered stone added for coloration, often with multiple colors worked in. The latter is a common means of making attractive washstands, washstand tops and shower stalls, with the skilled working of multiple colors resulting in simulated staining patterns as is often found in natural marble or travertine.

## Resin Casting

Resin casting is a method of plastic casting where a mold is filled with a liquid synthetic resin, which then hardens. It is primarily used for small-scale production like industrial prototypes and mixing with a curing agent (polymerization catalyst) at room temperature and normal pressure. The resins are named by analogy with plant resins, but are synthetic monomers for making polymer plastics. The socalled synthetic resins used include polystyrene resin, polyurethane resin, epoxy resin, unsaturated polyester resin, acrylic resin and silicone resin.

Epoxy resin has a lower viscosity than polyurethane resin; polyester resin also shrinks markedly while curing. Acrylic resin, in particular the methyl methacrylate type of synthetic resin, produces acrylic glass (also called PMMA, Lucite, Plexiglass), which is not a glass but a plastic polymer that is transparent, and very hard. It is suitable for embedding objects (such as, for example, acrylic trophies), for display purposes. Styrene is a similar liquid monomer at room temperature, which will also polymerize into clear glass-like polystyrene plastic, with addition of a suitable catalyst.

A flexible mold can be made of latex rubber, room temperature vulcanized silicone rubber or other similar materials at relatively low cost, but can only be used for a limited number of castings.

The simplest method is gravity casting where the resin is poured into the mold and pulled down into all the parts by gravity. When the two part resin is mixed air bubbles tend to be introduced into the liquid which can be removed in a vacuum chamber. The casting can also be done in a vacuum chamber (when using open molds) to either extract these bubbles, or in a pressure pot, to reduce their size to the point where they aren't visible. Pressure and/or centrifugal force can be used to help push the liquid resin into all details of the mold. The mold can also be vibrated to expel bubbles.

Each unit requires some amount of hands-on labor, making the final cost per unit produced fairly high. This is in contrast to injection molding where the initial cost of creating the metal mold is higher, but the mold can be used to produce a much higher number of units, resulting in a lower cost per unit.

## Collectibles and Models

Resin casting is used to produce collectible and customized toys and figures like designer toys, garage kits and ball-jointed dolls, as well as scale models, either individual parts or entire models of objects like trains, aircraft or ships. They are generally produced in small quantities, from the tens to a few hundred copies, compared to injection-molded plastic figures which are produced in many thousands. Resin casting is more labor intensive than injection molding, and the soft molds used are worn down by each cast. The low initial investment cost of resin casting means that individual hobbyists can produce small runs for their own use, such as customization, while companies can use it to produce small runs for public sale.

The creation of a toy or figure start with the traditional sculpting process where the artist designs a clay sculpture. Where appropriate, for example when making a garage kit, the sculpture is dissected into several parts like head, torso, arms and legs. A flexible mold made from room temperature vulcanized (RTVt) silicone rubber is made for each part.

After the mold has been made, a synthetic resin - commonly polyurethane mixed with a curing agent, is poured into each mold cavity. Mixing the two liquid parts causes an exothermic reaction which generates heat and within minutes causes the material to harden, yielding castings or copies in the shape of the mold into which it has been poured. The molds are commonly halfdivided (like the hollowed chocolate Easter eggs with candy inside) and a release agent may be used to make removal of the hardened/set resin from the mold easier. The hardened resin casting is removed from the flexible mold and allowed to cool.

Due to aggressive nature of most compounds used for casting and the high temperature of the reaction the mold gradually degrades and loses small details. Typically, a flexible mold will yield between 25 and 100 castings depending upon the size of the part, the intensity of the heat generated.

Depending on the type of product it may then be cut or sanded to remove any casting artefacts like sprues and seams. Some products are also assembled and painted, while some models and kits, which are intended for the consumer to assemble, are left unfinished.

The ability of RTV silicone molds to reproduce even the tiniest detail means that many of these low volume castings are of very high quality. Quality of both original masters and resin castings varies due to differences in creator's skill, as well as casting techniques.

## Casting Process Simulation

Casting process simulation uses numerical methods to calculate cast component quality considering mold filling, solidification and cooling, and provides a quantitative prediction of casting mechanical properties, thermal stresses and distortion. Simulation accurately describes a cast component's quality up-front before production starts. The casting rigging can be designed with respect to the required component properties. This has benefits beyond a reduction in preproduction sampling, as the precise layout of the complete casting system also leads to energy, material, and tooling savings.

The software supports the user in component design, the determination of melting practice and casting methoding through to pattern and mold making, heat treatment, and finishing. This saves costs along the entire casting manufacturing route.

Casting process simulation was initially developed at universities starting from the early '70s, mainly in Europe and in the U.S., and is regarded as the most important innovation in casting technology over the last 50 years. Since the late '80s, commercial programs (such as AutoCAST and MAGMA) are available which make it possible for foundries to gain new insight into what is happening inside the mold or die during the casting process.

## Sand Casting

Sand casting, also known as sand molded casting, is a metal casting process characterized by using sand as the mold material. The term "sand casting" can also refer to an object produced via the sand casting process. Sand castings are produced in specialized factories called foundries.

Over 70% of all metal castings are produced via a sand casting process.

Sand casting is relatively cheap and sufficiently refractory even for steel foundry use. In addition to the sand, a suitable bonding agent (usually clay) is mixed or occurs with the sand. The mixture is moistened, typically with water, but sometimes with other substances, to develop strength and plasticity of the clay and to make the aggregate suitable for molding. The sand is typically contained in a system of frames or mold boxes known as a flask. The mold cavities and gate system are created by compacting the sand around models, or patterns, or carved directly into the sand.

## Basic Process

There are six steps in this process:

1. Place a pattern in sand to create a mold.
2. Incorporate the pattern and sand in a gating system.
3. Remove the pattern.
4. Fill the mold cavity with molten metal.
5. Allow the metal to cool.
6. Break away the sand mold and remove the casting.

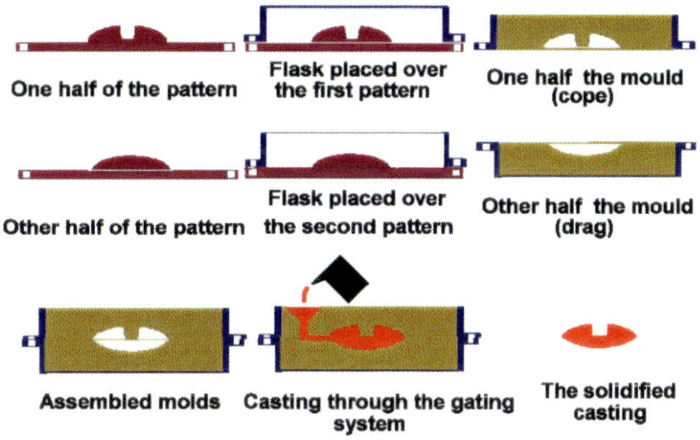


## Components

## Patterns

From the design, provided by an engineer or designer, a skilled pattern maker builds a pattern of the object to be produced, using wood, metal, or a plastic such as expanded polystyrene. Sand can be ground, swept or strickled into shape. The metal to be cast will contract during solidification, and this may be non-uniform due to uneven cooling. Therefore, the pattern must be slightly larger than the finished product, a difference known as contraction allowance.

Pattern-makers are able to produce suitable patterns using "contraction rules" (these are sometimes called "shrink allowance rulers" where the ruled markings are deliberately made to a larger spacing according to the percentage of extra length needed). Different scaled rules are used for different metals, because each metal and alloy contracts by an amount distinct from all others.

Patterns also have core prints that create registers within the molds into which are placed sand cores. Such cores, sometimes reinforced by wires, are used to create under-cut profiles and cavities which cannot be molded with the cope and drag, such as the interior passages of valves or cooling passages in engine blocks.

Paths for the entrance of metal into the mold cavity constitute the runner system and include the sprue, various feeders which maintain a good metal 'feed', and in-gates which attach the runner system to the casting cavity. Gas and steam generated during casting exit through the permeable sand or via risers, which are added either in the pattern itself, or as separate pieces.

## Molding Box and Materials

A multi-part molding box (known as a casting flask, the top and bottom halves of which are known respectively as the cope and drag) is prepared to receive the pattern. Molding boxes are made in segments that may be latched to each other and to end closures. For a simple object- flat on one side—the lower portion of the box, closed at the bottom, will be filled with a molding sand. The sand is packed in through a vibratory process called ramming, and in this case, periodically screeded level. The surface of the sand may then be stabilized with a sizing compound. The pattern is placed on the sand and another molding box segment is added.

Additional sand is rammed over and around the pattern. Finally a cover is placed on the box and it is turned and unlatched, so that the halves of the mold may be parted and the pattern with its sprue and vent patterns removed. Additional sizing may be added and any defects introduced by the removal of the pattern are corrected. The box is closed again. This forms a "green" mold which must be dried to receive the hot metal. If the mold is not sufficiently dried a steam explosion can occur that can throw molten metal about. In some cases, the sand may be oiled instead of moistened, which makes possible casting without waiting for the sand to dry. Sand may also be bonded by chemical binders, such as furane resins or amine-hardened resins.

## Chills

To control the solidification structure of the metal, it is possible to place metal plates, chills, in the mold. The associated rapid local cooling will form a finer-grained structure and may form a somewhat harder metal at these locations. In ferrous castings, the effect is similar to quenching metals in forge work. The inner diameter of an engine cylinder is made hard by a chilling core.

In other metals, chills may be used to promote directional solidification of the casting. In controlling the way a casting freezes, it is possible to prevent internal voids or porosity inside castings.

## Cores

To produce cavities within the casting-such as for liquid cooling in engine blocks and cylinder heads—negative forms are used to produce cores. Usually sand-molded, cores are inserted into the casting box after removal of the pattern. Whenever possible, designs are made that avoid the use of cores, due to the additional set-up time and thus greater cost.

With a completed mold at the appropriate moisture content, the box containing the sand mold is then positioned for filling with molten metal-typically iron, steel, bronze, brass, aluminium, magnesium alloys, or various pot metal alloys, which often include lead, tin, and zinc. After filling with liquid metal the box is set aside until the metal is sufficiently cool to be strong. The sand is then removed revealing a rough casting that, in the case of iron or steel, may still be glowing red. When casting with metals like iron or lead, which are significantly heavier than the casting sand, the casting flask is often covered with a heavy plate to prevent a problem known as floating the mold. Floating the mold occurs when the pressure of the metal pushes the sand above the mold cavity out of shape, causing the casting to fail.

After casting, the cores are broken up by rods or shot and removed from the casting. The metal from the sprue and risers is cut from the rough casting. Various heat treatments may be applied to relieve stresses from the initial cooling and to add hardness—in the case of steel or iron, by quenching in water or oil. The casting may be further strengthened by surface compression treatment—like shot peening—that adds resistance to tensile cracking and smooths the rough surface.

## Design Requirements

The part to be made and its pattern must be designed to accommodate each stage of the process, as it must be possible to remove the pattern without disturbing the molding sand and to have proper locations to receive and position the cores. A slight taper, known as draft, must be used on surfaces perpendicular to the parting line, in order to be able to remove the pattern from the mold. This requirement also applies to cores, as they must be removed from the core box in which they are formed. The sprue and risers must be arranged to allow a proper flow of metal and gasses within the mold in order to avoid an incomplete casting. Should a piece of core or mold become dislodged it may be embedded in the final casting, forming a sand pit, which may render the casting unusable. Gas pockets can cause internal voids. These may be immediately visible or may only be revealed after extensive machining has been performed. For critical applications, or where the cost of wasted effort is a factor, non-destructive testing methods may be applied before further work is performed.

## Process

In general, we can distinguish between two methods of sand casting: The first one using green sand and the second being the air set method.

## Green Sand

These expendable molds are made of wet sands that are used to make the mold's shape. The name comes from the fact that wet sands are used in the molding process. Green sand is not green in color, but "green" in the sense that it is used in a wet state (akin to green wood). Unlike the name suggests, "green sand" is not a type of sand on its own, but is rather a mixture of:

- Silica sand ($SiO_2$), or chromite sand ($FeCr_2O$), or zircon sand ($ZrSiO_4$), 75 to 85%, or olivine, or staurolite, or graphite.
- Bentonite (clay), 5 to 11%
- Water, 2 to 4%
- Inert sludge 3 to 5%

- Anthracite (0 to 1%)

There are many recipes for the proportion of clay, but they all strike different balances between moldability, surface finish, and ability of the hot molten metal to degas. The coal, typically referred to in foundries as sea-coal, which is present at a ratio of less than 5%, partially combusts in the presence of the molten metal leading to offgassing of organic vapors. Green sand for nonferrous metals do not use coal additives since the CO created is not effective to prevent oxidation. Green Sand for aluminum typically uses olivine sand (a mixture of the minerals forsterite and fayalite which are made by crushing dunite rock). The choice of sand has a lot to do with the temperature that the metal is poured. At the temperatures that copper and iron are poured, the clay gets inactivated by the heat in that the montmorillonite is converted to illite, which is a non-expanding clay. Most foundries do not have the very expensive equipment to remove the burned out clay and substitute new clay, so instead, those that pour iron typically work with silica sand that is inexpensive compared to the other sands. As the clay is burned out, newly mixed sand is added and some of the old sand is discarded or recycled into other uses. Silica is the least desirable of the sands since metamorphic grains of silica sand have a tendency to explode to form sub-micron sized particles when thermally shocked during pouring of the molds. These particles enter the air of the work area and can lead to silicosis in the workers. Iron foundries spend a considerable effort on aggressive dust collection to capture this fine silica. The sand also has the dimensional instability associated with the conversion of quartz from alpha quartz to beta quartz at 1250 degrees F. Often additives such as wood flour are added to create a space for the grains to expand without deforming the mold. Olivine, Chromite, etc. are used because they do not have a phase conversion that causes rapid expansion of the grains, as well as offering greater density, which cools the metal faster and produces finer grain structures in the metal. Since they are not metamorphic minerals, they do not have the polycrystals found in silica, and subsequently do not form hazardous sub-micron sized particles.

## Air Set Method

The air set method uses dry sand bonded with materials other than clay, using a fast curing adhesive. The latter may also be referred to as no bake mold casting. When these are used, they are collectively called "air set" sand castings to distinguish them from "green sand" castings. Two types of molding sand are natural bonded (bank sand) and synthetic (lake sand); the latter is generally preferred due to its more consistent composition.

With both methods, the sand mixture is packed around a pattern, forming a mold cavity. If necessary, a temporary plug is placed in the sand and touching the pattern in order to later form a channel into which the casting fluid can be poured. Air-set molds are often formed with the help of a casting flask having a top and bottom part, termed the cope and drag. The sand mixture is tamped down as it is added around the pattern, and the final mold assembly is sometimes vibrated to compact the sand and fill any unwanted voids in the mold. Then the pattern is removed along with the channel plug, leaving the mold cavity. The casting liquid (typically molten metal) is then poured into the mold cavity. After the metal has solidified and cooled, the casting is separated from the sand mold. There is typically no mold release agent, and the mold is generally destroyed in the removal process.

The accuracy of the casting is limited by the type of sand and the molding process. Sand castings made from coarse green sand impart a rough texture to the surface, and this makes them easy to identify. Castings made from fine green sand can shine as cast but are limited by the depth to width ratio of pockets in the pattern. Air-set molds can produce castings with smoother surfaces than coarse green sand but this method is primarily chosen when deep narrow pockets in the pattern are necessary, due to the expense of the plastic used in the process. Air-set castings can typically be easily identified by the burnt color on the surface. The castings are typically shot blasted to remove that burnt color. Surfaces can also be later ground and polished, for example when making a large bell. After molding, the casting is covered with a residue of oxides, silicates and other compounds. This residue can be removed by various means, such as grinding, or shot blasting.

During casting, some of the components of the sand mixture are lost in the thermal casting process. Green sand can be reused after adjusting its composition to replenish the lost moisture and additives. The pattern itself can be reused indefinitely to produce new sand molds. The sand molding process has been used for many centuries to produce castings manually. Since 1950, partially automated casting processes have been developed for production lines.

## Mold Materials

There are four main components for making a sand casting mold: base sand, a binder, additives, and a parting compound.

## Molding Sands

Molding sands, also known as foundry sands, are defined by eight characteristics: refractoriness, chemical inertness, permeability, surface finish, cohesiveness, flowability, collapsibility, and availability/cost.

Refractoriness — This refers to the sand's ability to withstand the temperature of the liquid metal being cast without breaking down. For example some sands only need to withstand 650°C (1,202° F) if casting aluminum alloys, whereas steel needs a sand that will withstand 1,500°C (2,730°F). Sand with too low a refractoriness will melt and fuse to the casting.

Chemical inertness — The sand must not react with the metal being cast. This is especially important with highly reactive metals, such as magnesium and titanium.

Permeability — This refers to the sand's ability to exhaust gases. This is important because during the pouring process many gases are produced, such as hydrogen, nitrogen, carbon dioxide, and steam, which must leave the mold otherwise casting defects, such as blow holes and gas holes, occur in the casting. Note that for each cubic centimeter (cc) of water added to the mold 16,000 cc of steam is produced.

Surface finish — The size and shape of the sand particles defines the best surface finish achievable, with finer particles producing a better finish. However, as the particles become finer (and surface finish improves) the permeability becomes worse.

Cohesiveness (or bond) — This is the ability of the sand to retain a given shape after the pattern is removed.

Flowability – The ability for the sand to flow into intricate details and tight corners without special processes or equipment.

Collapsibility — This is the ability of the sand to be easily stripped off the casting after it ha solidified. Sands with poor collapsibility will adhere strongly to the casting. When casting metals that contract a lot during cooling or with long freezing temperature ranges a sand with poor collapsibility will cause cracking and hot tears in the casting. Special additives can be used to improve collapsibility.

Availability/cost — The availability and cost of the sand is very important because for every ton of metal poured, three to six tons of sand is required. Although sand can be screened and reused, the particles eventually become too fine and require periodic replacement with fresh sand.

In large castings it is economical to use two different sands, because the majority of the sand will not be in contact with the casting, so it does not need any special properties. The sand that is in contact with the casting is called facing sand, and is designed for the casting on hand. This sand will be built up around the pattern to a thickness of 30 to 100 mm (1.2 to 3.9 in). The sand that fills in around the facing sand is called backing sand. This sand is simply silica sand with only a small amount of binder and no special additives.

## Plaster Mold Casting

Plaster mold casting is a metalworking casting process similar to sand casting except the molding material is plaster of paris instead of sand. Like sand casting, plaster mold casting is an expendable mold process, however it can only be used with non-ferrous materials. It is used for castings as small as 30 g (1 oz) to as large as 45 kg (99 lb). Generally, the form takes less than a week to prepare. Production rates of 1–10 units/hr can be achieved with plaster molds.

Parts that are typically made by plaster casting are lock components, gears, valves, fittings, tooling, and ornaments.

The plaster is not pure plaster of paris, but rather has additives to improve green strength, dry strength, permeability, and castability. For instance, talc or magnesium oxide are added to prevent cracking and reduce setting time; lime and cement limit expansion during baking; glass fibers increase strength; sand can be used as a filler. The ratio of ingredients is 70–80% gypsum and 20–30% additives.

The pattern is usually made from metal, however rubber molds may be used for complex geometry; these molds are called Rubber plaster molds. For example, if the casting includes reentrant angles or complex angular surfaces then the rubber is flexible enough to be removed, unlike metal. These molds are also inexpensive, reusable, more accurate than steel molds, fast to produce, and easy to change.

Typical tolerances are 0.1 mm (0.0039 in) for the first 50 mm (2.0 in) and 0.02 mm per additional centimeter (0.002 in per additional inch). A draft of 0.5 to 1 degree is required.

Standard surface finishes that are attainable are 1.3 to 4 micrometres (50–125 μin).

### Process

First, the plaster is mixed and the pattern is sprayed with a thin film of parting compound to prevent the plaster from sticking to the pattern. The plaster is then poured over the pattern and the unit shaken so that the plaster fills any small features. The plaster sets, usually in about 15 minutes, and the pattern is removed. The mold is then baked, between 120° C (248° F) and 260° C (500° F), to remove any excess water. The dried mold is then assembled, preheated, and the metal poured. Finally, after the metal has solidified, the plaster is broken from the cast part. The used plaster cannot be reused.

## Advantages and Disadvantages

Plaster mold casting is used when an excellent surface finish and good dimensional accuracy is required. Because the plaster has a low thermal conductivity and heat capacity the metal cools more slowly than in a sand mold, which allows the metal to fill thin cross-sections; the minimum possible cross-section is 0.6 mm (0.024 in). This results in a near net shape casting, which can be a cost advantage on complex parts. It also produces minimal scrap material.

The major disadvantage of the process is that it can only be used with lower melting temperature non-ferrous materials, such as aluminium, copper, magnesium, and zinc. The most commonly used materials are aluminium and copper. The maximum working temperature of plaster is 1,200° C (2,200° F), so higher melting temperature materials would melt the plaster mold. Also, the sulfur in the gypsum reacts with iron, making it unsuitable for casting ferrous materials.

Another disadvantage is that its long cooling times restrict production volume.

Plaster is not as stable as sand, so it is dependent on several factors, including the consistency of the plaster composition, pouring procedures, and curing techniques. If these factors are not closely monitored the mold can be distorted, shrink upon drying, have a poor surface finish, or fail completely.

Metalworking is the process of working with metals to create individual parts, assemblies, or large-scale structures. The term covers a wide range of work from large ships and bridges to precise engine parts and delicate jewelry. It therefore includes a correspondingly wide range of skills, processes, and tools.

Metalworking is a science, art, hobby, industry and trade. Its historical roots span cultures, civilizations, and millennia. Metalworking has evolved from the discovery of smelting various ores, producing malleable and ductile metal useful for tools and adornments. Modern metalworking processes, though diverse and specialized, can be categorized as forming, cutting, or joining processes. Today's machine shop includes a number of machine tools capable of creating a precise, useful work piece Plaster is a building material used for coating walls and ceilings. Plaster starts as a dry powder similar to mortar or cement and like those materials it is mixed with water to form a paste which liberates heat and then hardens. Unlike mortar and cement, plaster remains quite soft after setting, and can be easily manipulated with metal tools or even sandpaper. These characteristics make plaster suitable for a finishing, rather than a load-bearing material.

The term plaster can refer to gypsum plaster (also known as plaster of Paris), lime plaster, or cement plaster.

## Types:

### Gypsum Plaster (Plaster of Paris)

Gypsum plaster, or plaster of Paris, is produced by heating gypsum to about 300°F (150°C) $2CaSO_4{\cdot}2H_2O + \text{Heat} \rightarrow 2CaSO_4{\cdot}\tfrac{1}{2}H_2O + 3H_2O$ (released as steam).

When the dry plaster powder is mixed with water, it re-forms into gypsum. The setting of unmodified plaster starts about 10 minutes after mixing and is complete in about 45 minutes; but not fully set for 72 hours. If plaster or gypsum is heated above 392° F (200° C), anhydrite is formed, which will also re-form as gypsum if mixed with water.

A large gypsum deposit at Montmartre in Paris led "calcined gypsum" (roasted gypsum or gypsum plaster) to be commonly known as "plaster of Paris".

Plasterers often use gypsum to simulate the appearance of surfaces of wood, stone, or metal, on movie and theatrical sets for example. Nowadays, theatrical plasterers often use expanded polystyrene, although the job title remains unchanged.

Plaster of Paris can be used to impregnate gauze bandages to make a sculpting material called modroc. It is used similarly to clay, as it is easily shaped when wet, yet sets into a resilient and lightweight structure. This is the material that was (and sometimes still is) used to make classic plaster orthopedic casts to protect limbs with broken bones, the medical use having been partly inspired by the artistic use. Set modroc is an early example of a composite material.

Depending on the temperature and duration of the heating process, gypsum converts to the hemihydrate or an anhydrous form. Two polymorphs are known of the hemihydrate. The anhydrous calcium sulfate occurs in three forms, called anhydrite I, II, and III. Each form hydrates differently. The hemihydrate converts to the dihydrate. The anhydrite III converted via the hemihydrate, whereas anhydrite II is converted directly into dihydrate without forming intermediates.

### Lime Plaster

Lime plaster is a mixture of calcium hydroxide and sand (or other inert fillers). Carbon dioxide in the atmosphere causes the plaster to set by transforming the calcium hydroxide into calcium carbonate (limestone). Whitewash is based on the same chemistry.

To make lime plaster, limestone (calcium carbonate) is heated to produce quicklime (calcium oxide). Water is then added to produce slaked lime (calcium hydroxide), which is sold as a wet putty or a white powder. Additional water is added to form a paste prior to use. The paste may be stored in air-tight containers. Once exposed to the atmosphere, the calcium hydroxide turns back into calcium carbonate, causing the plaster to set.

Lime plaster was a common building material for wall surfaces in a process known as lath and plaster, whereby a series of wooden strips on a studwork frame was covered with a semi-dry plaster that hardened into a surface. The plaster used in most lath and plaster construction was mainly lime plaster, with a cure time of about a month. To stabilize the lime plaster during curing, small amounts of plaster of Paris were incorporated into the mix. Because plaster of Paris sets quickly, "retardants" were used to slow setting time enough to allow workers to mix large working quantities of lime putty plaster. A modern form of this method uses expanded metal mesh over wood or metal structures, which allows a great freedom of design as it is adaptable to both simple and compound curves. Today this building method has been partly replaced with drywall, also composed mostly of gypsum plaster. In both these methods a primary advantage of the material is that it is resistant to a fire within a room and so can assist in reducing or eliminating structural damage or destruction provided the fire is promptly extinguished.

Lime plaster is used for frescoes, where pigments, diluted in water, are applied to the still wet plaster. USA and Iran are the main plaster producers in the world.

## Cement Plaster

Cement plaster is a mixture of suitable plaster, sand, portland cement and water which is normally applied to masonry interiors and exteriors to achieve a smooth surface. Interior surfaces sometimes receive a final layer of gypsum plaster. Walls constructed with stock bricks are normally plastered while face brick walls are not plastered. Various cement-based plasters are also used as proprietary spray fireproofing products. These usually use vermiculite as lightweight aggregate. Heavy versions of such plasters are also in use for exterior fireproofing, to protect LPG vessels, pipe bridges and vessel skirts.

Cement plaster was first introduced in America around 1909 and was often called by the generic name adamamant plaster after a prominent manufacturer of the time. The advantages of cement plaster noted at that time were its strength, hardness, quick setting time and durability.

## Plasterwork

Plaster may also be used to create complex detailing for use in room interiors. These may be geometric (simulating wood or stone) or naturalistic (simulating leaves, vines, and flowers) These are also often used to simulate wood or stone detailing found in more substantial buildings.

## In Art

Many of the greatest mural paintings in Europe, like Michelangelo's Sistine Chapel ceiling are executed in fresco, meaning they are painted on a thin layer of wet plaster, called intonaco; the pigments sink into this layer so that the plaster itself becomes the medium holding them, which accounts for the excellent durability of fresco. Additional work may be added a secco on top of the dry plaster, though this is generally less durable.

Plaster may be cast directly into a damp clay mold. In creating this piece molds (molds designed for making multiple copies) or waste molds (for single use) would be made of plaster. This "negative" image, if properly designed, may be used to produce clay productions, which when fired in a kiln become terra cotta building decorations, or these may be used to create cast concrete sculptures. If a plaster positive was desired this would be constructed or cast to form a durable image artwork. As a model for stonecutters this would be sufficient. If intended for producing a bronze casting the plaster positive could be further worked to produce smooth surfaces. An advantage of this plaster image is that it is relatively cheap; should a patron approve of the durable image and be willing to bear further expense, subsequent molds could be made for the creation of a wax image to be used in lost wax casting, a far more expensive process. In lieu of producing a bronze image suitable for outdoor use the plaster image may be painted to resemble a metal image; such sculptures are suitable only for presentation in a weather-protected environment.

Plaster expands while hardening, then contracts slightly just before hardening completely. This makes plaster excellent for use in molds, and it is often used as an artistic material for casting.

Plaster is also commonly spread over an armature (form), usually made of wire, mesh or other materials, a process raised details. For these processes, limestone or acrylic based plaster may be employed.

## In Medicine

Plaster is widely used as a support for broken bones; a bandage impregnated with plaster is moistened and then wrapped around the damaged limb, setting into a close-fitting yet easily removed tube, known as an orthopedic cast.

Plaster is also used within radiotherapy when making immobilization casts for patients. Plaster bandages are used when constructing an impression of the patients head and neck, and liquid plaster is used to fill the impression and produce a plaster bust. Perspex is then vacuum formed over this bust creating an immobilization shell.

In dentistry, plaster is used for mounting casts or models of oral tissues. These diagnostic and working models are usually made from dental stone, a stronger, harder and denser derivative of plaster which is manufactured from gypsum under pressure. Plaster is also used to invest or flask wax dentures, the wax being subsequently removed and replaced with the final denture base material which is cured in the plaster mold.

In orthotics and prosthetics, plaster bandages traditionally where used to create impressions of the patients limb (or residuum). This negative impression was then, itself, filled with plaster of paris, to create a positive model of the limb and used in fabricating the final medial device.

In addition, dentures (false teeth) are made by first taking a dental impression using a soft, pliable material that can be removed from around the teeth and gums without loss of fidelity and using the impression to creating a wax model of the teeth and gums. The model is used to create a plaster mold (which is heated so the wax melts and flows out) and the denture materials are injected into the mold. After a curing period, the mold is opened and the dentures are cleaned up and polished.

## In Fire Protection

Plasters have been in use in passive fire protection, as fireproofing products, for many decades.

The finished plaster releases water vapor when exposed to flame, acting to slow the spread of the fire, for as much as an hour or two depending on thickness. It also provides some insulation to retard heat flow into structural steel elements, that would otherwise lose their strength and collapse in a fire. Early versions of these plasters have used asbestos fibres, which have by now been outlawed in industrialized nations and have caused significant removal and re-coating work. More modern plasters fall into the following categories.

Fibrous (including mineral wool and glass fiber) Cement mixtures either with mineral wool or with vermiculite Gypsum plasters, leavened with polystyrene beads, as well as chemical expansion agents to decrease the density of the finished product One differentiates between interior and exterior fireproofing. Interior products are typically less substantial, with lower densities and lower cost. Exterior products have to withstand more extreme fire and other environmental conditions. Exterior products are also more likely to be attractively tooled, whereas their interior cousins are usually merely sprayed in place. A rough surface is typically forgiven inside of buildings as dropped ceilings often hide them. Exterior fireproofing plasters are losing ground to more costly intumescent and endothermic products, simply on technical merit. Trade jurisdiction on unionized construction sites in North America remains with the plasterers, regardless of whether the plaster is decorative in nature or is used in passive fire protection. Cementitious and gypsum based plasters tend to be endothermic.

Fireproofing plasters are closely related to firestop mortars. Most firestop mortars can be sprayed and tooled very well, due to the fine detail work that is required of firestopping, which leads their mix designers to utilise concrete admixtures, that enable easier tooling than common mortars.

### In Industry

It is used in glass making, tanning, bleaching powder making and purification of sugar.

## Shell Molding

Shell molding, also known as shell-mold casting, is an expendable mold casting process that uses a resin covered sand to form the mold. As compared to sand casting, this process has better dimensional accuracy, a higher productivity rate, and lower labor requirements. It is used for small to medium parts that require high precision. Shell mold casting is a metal casting process similar to sand casting, in that molten metal is poured into an expendable mold. However, in shell mold casting, the mold is a thin-walled shell created from applying a sand-resin mixture around a pattern. The pattern, a metal piece in the shape of the desired part, is reused to form multiple shell molds. A reusable pattern allows for higher production rates, while the disposable molds enable complex geometries to be cast. Shell mold casting requires the use of a metal pattern, oven, sand-resin mixture, dump box, and molten metal.

Shell mold casting allows the use of both ferrous and non-ferrous metals, most commonly using cast iron, carbon steel, alloy steel, stainless steel, aluminum alloys, and copper alloys. Typical parts are small-to-medium in size and require high accuracy, such as gear housings, cylinder heads, connecting rods, and lever arms.

The shell mold casting process consists of the following steps:

Pattern creation - A two-piece metal pattern is created in the shape of the desired part, typically from iron or steel. Other materials are sometimes used, such as aluminum for low volume production or graphite for casting reactive materials.

Mold creation - First, each pattern half is heated to 175-370°C (350-700°F) and coated with a lubricant to facilitate removal. Next, the heated pattern is clamped to a dump box, which contains a mixture of sand and a resin binder. The dump box is inverted, allowing this sand-resin mixture to coat the pattern. The heated pattern partially cures the mixture, which now forms a shell around the pattern. Each pattern half and surrounding shell is cured to completion in an oven and then the shell is ejected from the pattern.

Mold assembly - The two shell halves are joined together and securely clamped to form the complete shell mold. If any cores are required, they are inserted prior to closing the mold. The shell mold is then placed into a flask and supported by a backing material.

Pouring - The mold is securely clamped together while the molten metal is poured from a ladle into the gating system and fills the mold cavity.

Cooling - After the mold has been filled, the molten metal is allowed to cool and solidify into the shape of the final casting.

Casting removal - After the molten metal has cooled, the mold can be broken and the casting removed. Trimming and cleaning processes are required to remove any excess metal from the feed system and any sand from the mold.

Examples of shell molded items include gear housings, cylinder heads and connecting rods. It is also used to make high-precision molding cores.

## Process

The process of creating a shell mold consists of six steps:

Fine silica sand that is covered in a thin (3–6%) thermosetting phenolic resin and liquid catalyst is dumped, blown, or shot onto a hot pattern. The pattern is usually made from cast iron and is heated to 230 to 315° C (450 to 600x° F). The sand is allowed to sit on the pattern for a few minutes to allow the sand to partially cure.

The pattern and sand are then inverted so the excess sand drops free of the pattern, leaving just the "shell". Depending on the time and temperature of the pattern the thickness of the shell is 10 to 20 mm (0.4 to 0.8 in).

The pattern and shell together are placed in an oven to finish curing the sand. The shell now has a tensile strength of 350 to 450 psi (2.4 to 3.1 MPa).

The hardened shell is then stripped from the pattern.

Two or more shells are then combined, via clamping or gluing using a thermoset adhesive, to form a mold. This finished mold can then be used immediately or stored almost indefinitely.

For casting the shell mold is placed inside a flask and surrounded with shot, sand, or gravel to reinforce the shell.

The machine that is used for this process is called a shell molding machine. It heats the pattern, applies the sand mixture, and bakes the shell.

Setup and production of shell mold patterns takes weeks, after which an output of 5–50 pieces/hr-mold is attainable. Common materials include cast iron, aluminum and copper alloys.

Aluminum and magnesium products average about 13.5 kg (30 lb) as a normal limit, but it is possible to cast items in the 45–90 kg (100–200 lb) range. The small end of the limit is 30 g (1 oz). Depending on the material, the thinnest cross-section castable is 1.5 to 6 mm (0.06 to 0.24 in). The minimum draft is 0.25 to 0.5 degrees.

Typical tolerances are 0.005 mm/mm or in/in because the sand compound is designed to barely shrink and a metal pattern is used. The cast surface finish is 0.3–4.0 micrometers (50–150 µin) because finer sand is used. The resin also assists in forming a very smooth surface. The process, in general, produces very consistent castings from one casting to the next.

The sand-resin mix can be recycled by burning off the resin at high temperatures.

## Advantages and Disadvantages

One of the greatest advantages of this process is that it can be completely automated for mass production. The high productivity, low labor costs, good surface finishes, and precision of the process can more than pay for itself if it reduces machining costs. There are also few problems due to gases, because of the absence of moisture in the shell, and the little gas that is still present easily escapes through the thin shell. When the metal is poured some of the resin binder burns out on the surface of the shell, which makes shaking out easy.

One disadvantage is that the gating system must be part of the pattern because the entire mold is formed from the pattern, which can be expensive. Another is the resin for the sand is expensive, however not much is required because only a shell is being formed.

Thin-walled: Complex, Solid: Cylindrical, Solid: Cubic, Solid: Complex (Flat, Thin-walled: Cylindrical, Thin-walled: Cubic) Part size Weight: 0.5 oz - 220 lb Weight: 1 oz - 450 ton Materials Metals, Alloy Steel, Carbon Steel, Cast Iron, Stainless Steel, Aluminum, Copper, Nickel Metals, Alloy Steel, Carbon Steel, Cast Iron, Stainless Steel, Aluminum, Copper, Magnesium, Nickel (Lead, Tin, Titanium, Zinc) Surface finish - Ra (µin) 50 - 300 (32 - 500) 300 - 600 (125 - 2000) Tolerance (in.) ± 0.015 (± 0.006) ± 0.03 (± 0.015) Max wall thickness 0.06 - 2.0 0.125 - 5 (0.09 - 40) Quantity 1000 - 1000000 (100 - 1000000) 1 - 1000 (1 - 1000000) Lead time Weeks (Days) Days (Hours).

Advantages: Can form complex shapes and fine details, Very good surface finish, High production rate, Low labor cost, Low tooling cost, Little scrap generated. Can produce very large parts, Can form complex shapes, Many material options, Low tooling and equipment cost, Scrap can be recycled, Short lead time possible.

Disadvantages: High equipment cost, Poor material strength, High porosity possible, Poor surface finish and tolerance, Secondary machining often required, Low production rate, High labor cost.

Applications: Cylinder heads, connecting rods Engine blocks and manifolds, machine bases, gears, pulleys.

Resin in the most specific use of the term is a hydrocarbon secretion of many plants, particularly coniferous trees. Resins are valued for their chemical properties and associated uses, such as the production of varnishes, adhesives and food glazing agents. They are also prized as an important source of raw materials for organic synthesis, and as constituents of incense and perfume. Plant resins have a very long history that was documented in ancient Greece by Theophrastus, in ancient Rome by Pliny the Elder, and especially in the resins known as frankincense and myrrh, prized in ancient Egypt. These were highly prized substances, and required as incense in some religious rites. Amber is a hard fossilized resin from ancient trees.

More broadly, the term "resin" also encompasses a great many synthetic substances of similar mechanical properties (thick liquids that harden into transparent solids), as well as shellacs of insects of the superfamily Coccoidea.

Other liquid compounds found inside plants or exuded by plants, such as sap, latex, or mucilage, are sometimes confused with resin, but are not chemically the same. Saps, in particular, serve a nutritive function that resins do not. There is no consensus on why plants secrete resins.

However, resins consist primarily of secondary metabolites or compounds that apparently play no role in the primary physiology of a plant. While some scientists view resins only as waste products, their protective benefits to the plant are widely documented. The toxic resinous compounds may confound a wide range of herbivores, insects, and pathogens; while the volatile phenolic compounds may attract benefactors such as parasitoids or predators of the herbivores that attack the plant.

The word "resin" has been applied in the modern world to nearly any component of a liquid that will set into a hard lacquer or enamel-like finish. An example is nail polish, a modern product which contains "resins" that are organic compounds, but not classical plant resins. Certain "casting resins" and synthetic resins (such as epoxy resin) have also been given the name "resin" because they solidify in the same way as some plant resins, but synthetic resins are liquid monomers of thermosetting plastics, and do not derive from plants.

## Resin of a Pine

The resin produced by most plants is a viscous liquid, composed mainly of volatile fluid terpenes, with lesser components of dissolved non-volatile solids which make resin thick and sticky. The most common terpenes in resin are the bicyclic terpenes alpha-pinene, beta-pinene, delta-3 carene and sabinene, the monocyclic terpenes limonene and terpinolene, and smaller amounts of the tricyclic sesquiterpenes, longifolene, caryophyllene and delta-cadinene. Some resins also contain a high proportion of resin acids. The individual components of resin can be separated by fractional distillation.

A few plants produce resins with different compositions, most notably Jeffrey Pine and Gray Pine, the volatile components of which are largely pure n-heptane with little or no terpenes. The exceptional purity of the n-heptane distilled from Jeffrey Pine resin, unmixed with other isomers of heptane, led to its being used as the defining zero point on the octane rating scale of petrol quality. Because heptane is highly flammable, distillation of resins containing it is very dangerous. Some resin distilleries in California exploded because they mistook Jeffrey Pine for the similar but terpene-producing Ponderosa Pine. At the time the two pines were considered to be the same species of pine; they were only classified as separate species in 1853.

Some resins when soft are known as 'oleoresins', and when containing benzoic acid or cinnamic acid they are called balsams. Oleoresins are naturally occurring mixtures of an oil and a resin; they can be extracted from various plants. Other resinous products in their natural condition are a mix with gum or mucilaginous substances and known as gum resins. Many compound resins have distinct and characteristic odors, from their admixture with essential oils.

Certain resins are obtained in a fossilized condition, amber being the most notable instance of this class; African copal and the kauri gum of New Zealand are also procured in a semi-fossil condition.

Solidified resin from which the volatile terpene components have been removed by distillation is known as rosin. Typical rosin is a transparent or translucent mass, with a vitreous fracture and a faintly yellow or brown colour, non-odorous or having only a slight turpentine odour and taste.

It is insoluble in water, mostly soluble in alcohol, essential oils, ether and hot fatty oils, and softens and melts under the influence of heat, is not capable of sublimation, and burns with a bright but smoky flame.

This comprises a complex mixture of different substances including organic acids named the resin acids. These are closely related to the terpenes, and derive from them through partial oxidation. Resin acids can be dissolved in alkalis to form resin soaps, from which the purified resin acids are regenerated by treatment with acids. Examples of resin acids are abietic acid (sylvic acid), $C_{20}H_{30}O_2$, plicatic acid contained in cedar, and pimaric acid, $C_{20}H_{30}O_2$, a constituent of galipot resin. Abietic acid can also be extracted from rosin by means of hot alcohol; it crystallizes in leaflets, and on oxidation yields trimellitic acid, isophthalic acid and terebic acid. Pimaric acid closely resembles abietic acid into which it passes when distilled in a vacuum; it has been supposed to consist of three isomers.

## Uses

The hard transparent resins, such as the copals, dammars, mastic and sandarac, are principally used for varnishes and adhesives, while the softer odoriferous oleo-resins (frankincense, elemi, turpentine, copaiba) and gum resins containing essential oils (ammoniacum, asafoetida, gamboge, myrrh, and scammony) are more largely used for therapeutic purposes and incense.

Resin in the form of rosin is applied to the bows of musical string instruments because of its ability to add friction to the hair to increase sound quality.

Ballet dancers, as well as boxers in the old days, may apply crushed resin to their shoes to increase grip on a slippery floor.

Resin has also been used as a medium for sculpture by artists such as Eva Hesse, and in other types of artwork.

In the early 1990s, most ten-pin bowling ball manufacturers started adding resin particles to the covers of bowling balls. Resin makes a bowling ball tackier than it would otherwise be, increasing its ability to hook into the pins at an angle and (with correct technique) making strikes easier to achieve.

# Evaporative-pattern Casting

The evaporative pattern casting is one of the modern and hybrid casting processes, which aims to minimize the production cost, and is capable of producing intricate shapes with less efforts in pattern making.

The evaporative pattern casting process is also known by other names such as full mold casting process and lost foam casting process. It makes use of patterns made from evaporative materials like foam.

The pattern material gets evaporated due to the heat of molten metal as it gets filled. The foam patterns were first patented by H F Shroyer, for metal castings way back in the year 1958. Earlier expanded polystyrene also known as EPS blocks were used for measuring the patterns, bonded sand was used for support while pouring.

The process is called as full mold process wherein the pattern is machined from an EPS block. The process is primarily used to make large one of castings.

In the year 1964, M C Flemming used unbounded sand and the process started getting recognized, as lost foam casting process, also known as LFC process. In the LFC the foam pattern is molded from polystyrene beads. In full mold process on the other hand, bonded sand is used. In the LFC process, the bonded sand is replaced by unbounded sand. The evaporative pattern casting process is a binder less process, wherein no physical bonding is required to bind a sand particles.

Foam casting techniques have been known by a variety of generic and proprietary names such as, lost foam casting, evaporative pattern casting, cavity less casting, full mold casting and evaporative foam casting. In foam castings evaporative polystyrene is used to prepare the complete pattern, including the gates and risers.

The pattern is further embedded in a no bake type sand. The molten metal is poured through the sprue while the pattern is still inside the mold. The heat of the molten metal is sufficient enough to gasify the pattern and the pattern progressively gates displaced by it. Using unbounded sand and expandable polystyrene pattern, the EPC process is a very economical method in producing complex and closed tolerance castings.

The expandable polystyrene can be molded into numerous complex and rigid shapes and it is basically a thermoplastic material. In the EPC process expandable polystyrene patterns are attached to an expandable polystyrene gating system. A refractory coating is applied to the entire assembly, once the coating gets dried, the entire foam pattern assembly is kept on loose dry sand in a vented flask. The flask is then vibrated and additional sand is added to it until the pattern assembly gets completely embedded in sand. Subsequently, molten metal is poured into the sprue, which further reprises the foam polystyrene. As soon as it enters inside the entered metal solidifies and reproduces the size and shape of the used pattern perfectly to get the desired casting in this process.

The pattern used refers to an expandable polystyrene or foamed polystyrene part, which gets vaporized by the molten metal. Therefore, for every casting process a new pattern is required to be made, which eventually gets evaporated. Now, let us see the advantages of the evaporative pattern casting process.

## Advantages of EPC Process

In the evaporative pattern casting process no cores are required, making it the most advantageous process. No requirement for binders or other additives, as it is a binder less process. Complete sand reclamation is possible using very simple and inexpensive techniques. In this process sand shake out is easy as the sand is unbounded. Since, the pattern used in EPC process is one piece, hence no parting lines are required. Since, cores are eliminated, no core prints are required. Further due to this, there is no mismatch and problems due to the core sips are also reduced. This reduces the casting defects considerably improved casting quality is achieved in this process.

Close tolerances are possible the EPC is an environmentally favorable process. The EPC being a binder less process, efforts on cleaning the molded sand are virtually nil. The EPC process is viewed as a value added process, harder than a substitute for sand casting. The EPC is cheaper and much simpler than the investment casting process, as the foam is more readily available and cheaper as compared to wax.

The EPC process starts with burning of the foam from the edges and has a natural directional solidification phenomenon, thereby not requiring risers. In directional solidification, the solidification starts from the edges of the inner most sections and ends towards the top most portion. Now, let us also see the disadvantages or limitations of the evaporative pattern casting process.

## Disadvantages of EPC Process

Since, every casting requires a new pattern, it increases the cost. The initial cost for die making is also high. In low production items the costs further increases. There is a limitation on the minimum section thickness of the pattern. Due to lower strains of the patterns, they can get easily damaged or distorted. Quality of the casting fully depends upon the quality of the pattern.

During pouring as the sand is unbounded, due to the difference in the evaporation rates of the material and flow rate of the molten metal some of the sand may fall down in the cavities generated, thereby leading to a defective casting. Now, let us also see the applications of the evaporative pattern casting process.

## Applications

This process is used for making automotive components like cylinder heads, engine blocks, inlet manifolds, heat exchanger, crank shafts etcetera. It is also used for parts fabrication in marine, aerospace and construction industries. Now, let us see the evaporative pattern casting process.

## EPC Process

The process of evaporative pattern casting starts with the pre expansion of polystyrene beads. Once these pre expanded beads get stabilized, pattern sections are formed by blowing them into a mold. A cycle of steam is used to fully expand and make the beads fuse in the mold itself clusters are made.

Through assembly of the pattern sections using glue, the gating systems are also similarly, glued and attached. A ceramic coating is used to cover the foam cluster. The coating acts as a barrier preventing the penetration and sand erosion during pouring. Once the coating gets dried, the cluster is placed in a flask along with backing of the bonded sand. Any suitable sand can be used as long as it resists the temperature of the molten metal being poured. Silica sand zircon sand olivine's sand and chromites are commonly used molding sands.

The EPC process has high degree of sand re-clamation, hence one can use zircon or chromites sands which are otherwise considered expensive. The strength of the mold can be determined through the frictional resistance between the sand grains. Although higher bulk density is provided by the rounded grains, the mold strength which angular grains is much higher. In order to ensure uniform and proper compaction, a vibrating table is used for the mold compaction. After completing this process, the cluster is packed in a flask and the mold is set ready for getting poured. A schematic sketch of the EPC process is shown in the figure.

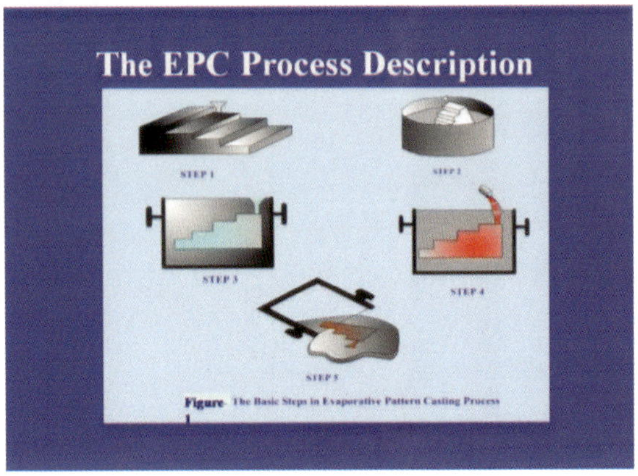


This is the pattern, the pattern is coated and then finally, the pouring is done, and the product is obtained. The patterns used in the EPC process are made of expandable polystyrene. A typical pattern is shown in the figure two, pattern fabrication starts with the pre expansion of polystyrene beads. They are blown into the die cavity to form pattern sections.

Figure: A polystyrene pattern

This is a typical polystyrene pattern. When the beads are in the die, a steam cycle causes them to fully expand and fuse together, forming the pattern as shown here. Ford fabrication of steamed pattern die and sprue a cavity, as per the pattern dimensions is prepared.

Aluminum is commonly used as the die material. The complete die is fabricated in two parts, called as male and female.

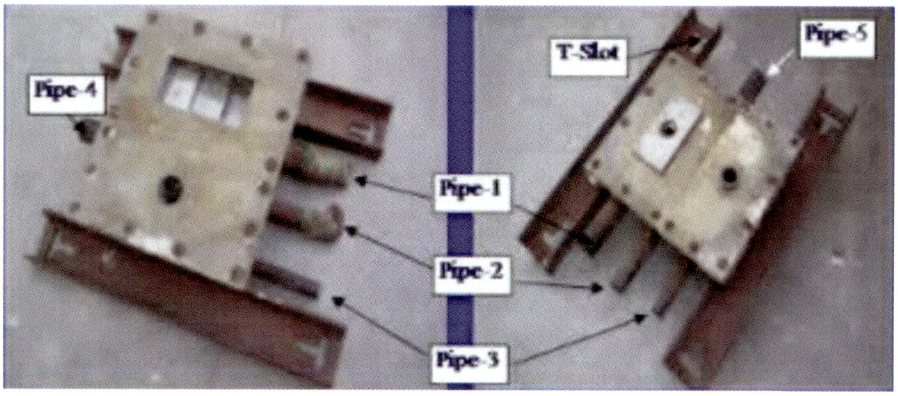


The three inlet pipes 1, 2 and 3 and one outlet pipe pipe 4 are attached with male and female parts, as shown in this figure. These are inlets this is the outlet. The function of all three inlet pipes was same in male and female parts. The two outermost pipes, number 1 and 3 were used to supply water as a cooling agent in the die. The middle pipe, which is pipe number 2 was used to supply the steam into the cavity through pin holes to soften and expand the beads.

Pipe 4 was used to throw water and steam from the die. Polystyrene beads were fed in to the die cavity through a pipe, which is connected at the top of the male part at the four corners of the die. These slots are fabricated to hold both male and female parts of the die in the phrase. Pattern density and bead size density and bead size play an important role in evaporative pattern casting process. In order to minimize the amount of gases evolved during evaporation of the pattern patterns with low density are generally used.

The low density patterns ensure that the gases escape out through the coatings, sand and vent into the atmosphere. Gas formation is a function of the metal pouring temperature and pattern density. If the gas forms faster than it can vent, a defective casting will be formed. If the pattern density is increased, then at a constant pouring temperature more gas is formed.

If the pattern density is held constant and the pouring temperature is increased more gas will be formed as the polystyrene molecules break into many more basic molecules. The gases formed should pass through the coating on the pattern surface. The pattern density generally varies between 1 and 1.5 PCF, which further depends on the geometry and metal, that is being poured. Small beads are required for obtaining relativity non beady smooth surfaces on the molded patterns.

Requirements of refractory coating: The coatings are an inherent part of the casting process and also provide a very good quality surface to the casting. Some specific demands as mentioned below are desired for a refractory coatings in the application of the EPC process. Highly permeable coating is preferred for rougher sand while medium and low permeable coating is preferred for final sand. Quick drying is preferred coating should get easily stuck to the pattern and there should be provision of controlling and adjusting coating layer thickness.

Appropriate strength resistance to abrasion, resistance to cracks during storage, resistance to bending and deformation during mold making are required. Thicker refractory coating layers are desired in cases where rough sands are used for molding at high temperatures. Coating material, the refractory coating materials play an intimate role in the EPC process. Its function is to provide a refractory protection and seal in penetration of metals.

The coating materials should provide smooth surfaces and act as a membrane between the polystyrene pattern and the unbounded sand mass. The unbounded sand mass should be firmly held in place, till the polystyrene foam pattern gets decomposed under the effect of the entering molten metal stream. This is achieved by allowing the permeability of the refractory coating to restrict the free flow of decomposition gases and liquids through the coating into the unbounded sand mass. This causes a back pressure against the entering molten metal stream resulting in a pressure force, that holds the unbounded sand in place.

The control of escaping gases through the sand mass is done through appropriate selection of the coating material and its permeability characteristics. The coatings are engineered to obtain a variety of properties imperative to the casting process such as thermal insulation, abrasion resistance and liquid absorption.

There are several types of refractory coatings available to be used on evaporative pattern. Some of the coating materials are are zircon sand, kaolin, talc etcetera. The coating generally recommended, in case of cast iron is iron powder, as it prevents the metal penetration. The siliminate quartz and aluminum silicate can be used as filler materials to make the coatings cost effective. Now, let us look into the process variables involved in the evaporative pattern casting process.

The process variables have a major effect on the quality of evaporative pattern casting process mold and castings. An Ishikawa cause and effect diagram is shown in the figure four.

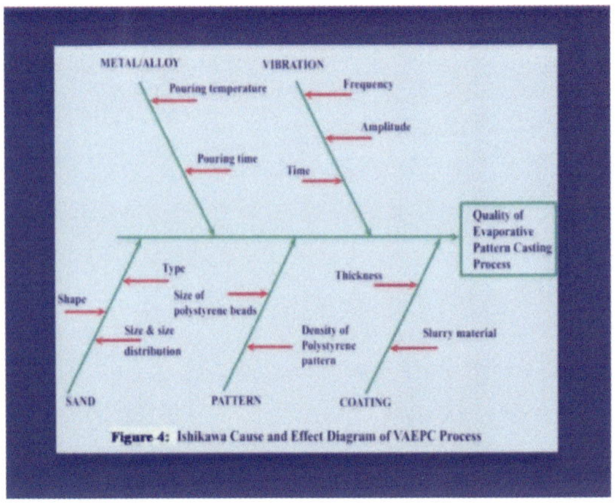


**Figure 4:** Ishikawa Cause and Effect Diagram of VAEPC Process

Here the major parameters involves the metals and alloys, the vibration, the sand used, the pattern and the coating. The parameters involved in this, affect ultimately the casting to be obtained as the product. The EPC process variables can be grouped in the following categories.

Number one, variables based on molding sand used, these are type of sand, shape of the sand grains, grain size and size distribution. Number two, variables based on pattern.

They are density and size of pattern beads. Number three, variables based on pattern coating the variables are, slurry material, base, filler and additives and coating thickness. The fourth category of variables are based on the vibration system.

The variables are, amplitude of vibration, time of vibration and the frequency of vibration. The fifth category of variables are based on, pouring, the variables are pouring time and pouring temperature. These are some of the product produced through evaporative pattern casting process.

## Some Product Example of EPC Process

The products obtained from evaporative pattern casting need to be subjected to various tests to ensure its soundness.

The most widely used technique for measuring the effect of temperature on the samples are thermo gravimetry also called TG test. Then derivative thermo gravimetry also called DTG test and differential thermal analysis called DTA test.

Then another very popular and very widely used technique is the X-ray diffraction analysis. This is quiet old process, but quiet comprehensive and quiet popular process. This is used to analyze the different phases present in a specimen. X-ray diffraction also gives information about the elements present and its composition. As we know different materials diffracts the light incident on it or we can say different electromagnetic radiation incident on heat depending on the structure. Therefore, this can be explained like this.

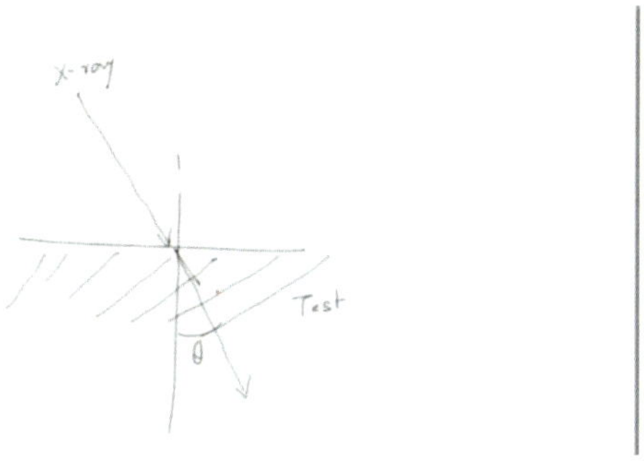


If this is the surface and if X-rays are falling on it, this is say X-rays, then this refracted x- rays inside this material, so this is the material, this is test material, this will be depending on the crystal structure of this material. Therefore, by measuring the angle of deflection, what is this angle? We can find out, what is the material in this because this can be correlated to its characteristic. That means this diffraction will be characteristic of the material or the crystal structure through which, X-ray incident X-ray is passing through.

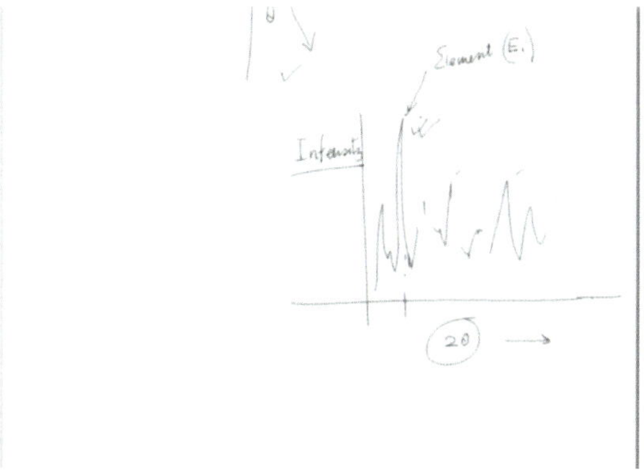


Therefore, we can get a signal like this. So, here so this is 2 theta and this is called intensity. Now, these different peaks, these peaks represents that one element some element say E 1 is present and that can be confirmed by subsequent detection of five peaks, five such peaks in this spectrum. This intensity is plotted as a function of 2 theta as we have indicated.

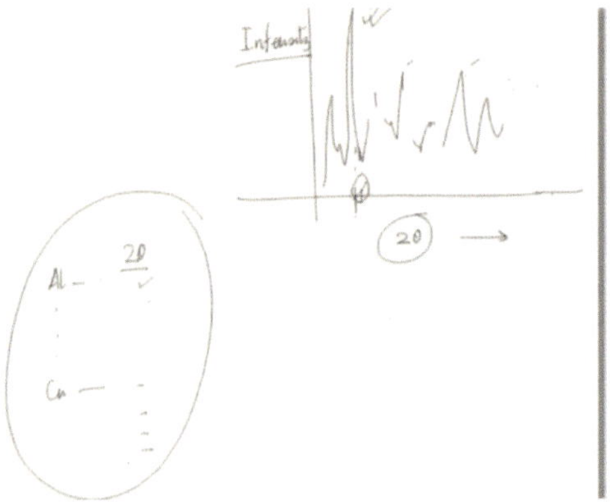


And by knowing this 2 theta, we can identify which element it is because this 2 theta for different materials say aluminum for different phases, this 2 theta is fixed. Similarly, for copper different phases, this value is fixed. Now, by correlating these values, if we obtain these values here, then by identifying the peak corresponding to that particular 2 theta value, one can identify the element present in that material. So, this is a very important technique widely used technique, a very low cost technique and of course, this is a non destructive testing technique, which is very widely used even today under crystallographic labs. Of course for subjecting this particular test to a material, the material needs to be crystalline in nature. Amorphous materials like glass etcetera cannot be characterized by using this method.

## Lost-foam Casting

Lost foam casting technique is known by different generic and propriety names like lost foam, evaporative pattern casting, evaporative foam casting, full mold, etc. Similar to the full mold process, in this process the pattern evaporates when the metal is poured into the mold.

Originally, full mold process was referred to as lost foam Casting. But, later the two processes were clearly differentiated by the use unboned dry silica in LFC while bonded sand was used in full mold process. This process gained world famous, when General Motors first implemented it to manufacture 4.3L, V-6 diesel cylinder head, in 1981.

Process: Lost foam casting is a type of metal casting process that uses expendable foam patterns to produce castings. Expanded polystyrene foam is used which melts when molten metal is poured into the mold.

Firstly, a pattern is made from form using a variety of methods. For small volume runs the pattern can be hand cut or machined from a solid block of foam, if the designs are simple enough it can even be cut using a hot-wire foam cutter.

If the volume is large, then the pattern can be mass-produced by a process similar to injection molding. Pre-expanded beads of polystyrene are injected or blown into a preheated aluminum mold at low pressure. Injecting steam into the tooling cavity expands the beads. The beads will flatten against the tool surface and stick to one another. After cooling, the pattern is ejected from the tool. The final pattern is approximately 97.5% air and 2.5% polystyrene. Secondly, pouring basins, runners, and risers can be hot glued to the patter to form the final pattern assembly. The assembly is then coated with refractory coatings, which prevents the sand mold from breaking during the casting process.

Finally, when the coating dries, the cluster is placed into a flask and surrounded with un-bonded sand. Sand is compacted around the pattern by vibrating the table. Molten metal is poured into this tight mold either mechanically or using automatic pouring machine. No flash is formed, since there is no draft or parting line is used.

Application: Lost foam casting is used mostly for automotive applications. Cast iron, aluminum alloys, steels, nickel and in some cases stainless steel and copper alloys are cast in this process. The flexibility of LFC is useful in making complicated casting assemblies for automotive parts like cylinder heads, weldments, etc. This simple and inexpensive method is used in hobby foundry work.

## Advantage:

- A simple and affordable technique that can be used in home foundries
- Cleaning cost is reduced in lost foam casing
- Tooling used for the casting process does not wear away
- Complex shapes can be cast in near net shapes
- A simple and inexpensive method for casting complex constructions
- Difficult designs can be done which in unattainable using other casting techniques

- Castings requires minimal machining
- Castings are dimensionally accurate with excellent surface finish.

### Full-mold Casting

Full-mold casting technique is quite similar to investment casting, but here polystyrene foam is used as the pattern. The foam pattern is laced with a refractory material. The pattern is put in an encasing of one-piece sand mold, during the pouring of the metal. The foam vaporizes and its place is taken by the metal.

- Complex shaped castings: Possible without any draft or flash.
- Minimum wall thicknesses: 2.5 mm
- Tolerances: Can be held to 3% on dimensions.
- Surface finish: From 2.5μm to 25μm (0.1μin to 1.0 μin) rms.
- Size limits: From 400 g (1 lb) to several tons.
- Draft allowance: Required.

Materials cast with this process are aluminum, iron, steels, nickel alloys, copper alloys to produce pump housings, manifolds, and auto brake components.

## Permanent Mold Casting

Permanent mold casting is a metal casting process that shares similarities to both sand casting and die casting. As in sand casting, molten metal is poured into a mold which is clamped shut until the material cools and solidifies into the desired part shape. However, sand casting uses an expendable mold which is destroyed after each cycle. Permanent mold casting, like die casting, uses a metal mold (die) that is typically made from steel or cast iron and can be reused for several thousand cycles. Because the molten metal is poured into the die and not forcibly injected, permanent mold casting is often referred to as gravity die casting.

Permanent mold casting is typically used for high-volume production of small, simple metal parts with uniform wall thickness. Non-ferrous metals are typically used in this process, such as aluminum alloys, magnesium alloys, and copper alloys. However, irons and steels can also be cast using graphite molds. Common permanent mold parts include gears and gear housings, pipe fittings, and other automotive and aircraft components such as pistons, impellers, and wheels.

The permanent mold casting process consists of the following steps:

1. Mold preparation - First, the mold is pre-heated to around 300-500°F (150-260°C) to allow better metal flow and reduce defects. Then, a ceramic coating is applied to the mold cavity surfaces to facilitate part removal and increase the mold lifetime.

2. Mold assembly - The mold consists of at least two parts - the two mold halves and any cores used to form complex features. Such cores are typically made from iron or steel, but expendable sand cores are sometimes used. In this step, the cores are inserted and the mold halves are clamped together.

3. Pouring - The molten metal is poured at a slow rate from a ladle into the mold through a sprue at the top of the mold. The metal flows through a runner system and enters the mold cavity.

4. Cooling - The molten metal is allowed to cool and solidify in the mold.

5. Mold opening - After the metal has solidified, the two mold halves are opened and the casting is removed.

6. Trimming - During cooling, the metal in the runner system and sprue solidify attached to the casting. This excess material is now cut away.

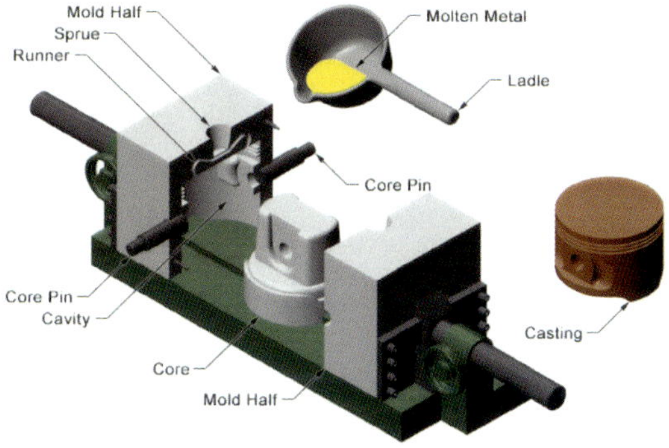


Figure: Permanent Mold Casting

Using these basic steps, other variations on permanent mold casting have been developed to accommodate specific applications. Examples of these variations include the following:

- Slush casting - As in permanent mold casting, the molten metal is poured into the mold and begins to solidify at the cavity surface. When the amount of solidified material is equal to the desired wall thickness, the remaining slush (material that has yet to completely solidify) is poured out of the mold. As a result, slush casting is used to produce hollow parts without the use of cores.

- Low pressure permanent mold casting - Instead of being poured, the molten metal is forced into the mold by low pressure air (< 1 bar). The application of pressure allows the mold to remain filled and reduces shrinkage during cooling. Also, finer details and thinner walls can be molded.

- Vacuum permanent mold casting - Similar to low pressure casting, but vacuum pressure is used to fill the mold. As a result, finer details and thin walls can be molded and the mechanical properties of the castings are improved

- Smooth filling of molds – To reduce risks of defects caused by inclusions and turbulence, which would compromise the product you receive, we use a grating design for reliably smooth mold fillings.

## Advantages of Permanent Mold Casting

- Superior physical attributes – Exceptional mechanical properties. Dominant casting strength is achieved by using the permanent mold casting process rather than the sand casting process.
- Excellent surface finishes – As cast products have top-quality finishes and dense surface structure, which makes them easier to polish.
- Closer net shaped profiles – Intricate shapes and designs can be cast by our expert team, as per your specifications, using the permanent mold process.
- Precise, consistent control – Our electric melting furnaces produce consistent metal heats while our permanent mold dies have precise dimension control, and our CNC machines work with unmatched precision.
- Weight reduction – As compared to the sand castings, the permanent mold castings are more lightweight due to both the materials used and the slimmer sections (as thin as 0.09”) held within them.
- Tight pressure – Get superior pressure tightness and soundness by using our high-quality permanent mold casting process for your order.
- Few defects or part variation – It is crucial that you get uniform parts for your projects, so we use metal dies that minimize part variation to as little at only +/-0.002”. Metal dies vary significantly less than sand-casting, which lowers part-to-part variability significantly.
- Tiny cast holes – Small holes (as tiny as 0.187”) can be cast into parts, with very close tolerances. This ability often eliminates the need for any secondary drilling and gets your finished part completed quickly.
- Eco-friendly – Bunting recycles grates and scrap through our furnaces to reduce our company’s carbon footprint. We believe in taking responsibility to minimize our impact on the environment and therefore support green practices.

## Limitations

The limitations of casting using the permanent mold or gravity process includes:

- Only materials with lower melting temperatures can be used in the process.
- While cast iron parts are manufactured using this process, the high melting point of cast iron does have a negative (or ‘wearing’) effect on the mold used.
- Part geometry is limited with this process as the mold needs to open and close.
- This particular casting process becomes a less viable option when compared to sand casting in cases where three or more sand cores are required.

Other limitations and disadvantages of casting using this particular process includes:

- A high tooling cost and a long set-up time.
- The permanent mold casting process is, in general, limited to smaller castings.
- Because of the high tooling cost involved, a high production volume is needed in order to make this process and economically viable manufacturing option.
- The higher the pouring temperature of the molten metal, the shorter the life of the mold.

## Die Casting

Die casting is a manufacturing process for producing accurately dimensioned, sharply defined, smooth or textured-surface metal parts. It is accomplished by forcing molten metal under high pressure into reusable metal dies. The process is often described as the shortest distance between raw material and finished product. The term "die casting" is also used to describe the finished part.

The term "gravity die casting" refers to castings made in metal molds under a gravity head.

### Producing Die Castings

First, a steel mold capable of producing tens of thousands of castings in rapid succession must be made in at least two sections to permit removal of castings. These sections are mounted securely in a machine and are arranged so that one is stationary (fixed die half) while the other is moveable (injector die half). To begin the casting cycle, the two die halves are clamped tightly together by the die casting machine. Molten metal is injected into the die cavity where it solidifies quickly. The die halves are drawn apart and the casting is ejected. Die casting dies can be simple or complex, having moveable slides, cores, or other sections depending on the complexity of the casting.

The complete cycle of the die casting process is by far the fastest known for producing precise non-ferrous metal parts. This is in marked contrast to sand casting which requires a new sand mold for each casting. While the permanent mold process uses iron or steel molds instead of sand, it is considerably slower, and not as precise as die casting.

### Types of Machines for Die Casting

Regardless of the type of machine used, it is essential that die halves, cores and/or other moveable sections be securely locked in place during the casting cycle. Generally, the clamping force of the machine is governed by (a) the projected surface area of the casting (measured at the die parting line) and (b) the pressure used to inject metal into the die. Most machines use toggle type mechanisms actuated by hydraulic cylinders (sometimes air pressure) to achieve locking. Others use direct acting hydraulic pressure. Safety interlock systems are used to prevent the die from opening during the casting cycles.

Die casting machines, large or small, vary fundamentally only in the method used to inject molten metal into the die. These are classified and described as either hot or cold chamber die casting machines.

## Hot Chamber Machines

Hot chamber machines are used primarily for zinc, and low melting point alloys which do not readily attack and erode metal pots, cylinders and plungers. Advanced technology and development of new, higher temperature materials has extended the use of this equipment for magnesium alloys.

In the hot chamber machine, the injection mechanism is immersed in molten metal in a furnace attached to the machine. As the plunger is raised, a port opens allowing molten metal to fill the cylinder. As the plunger moves downward sealing the port, it forces molten metal through the gooseneck and nozzle into the die. After the metal has solidified, the plunger is withdrawn, the die opens, and the resulting casting is ejected.

Hot chamber machines are rapid in operation. Cycle times vary from less than one second for small components weighing less than one ounce, to thirty seconds for a casting of several pounds. Dies are filled quickly (normally between five and forty milliseconds) and metal is injected at high pressures (1,500 to over 4,500 psi). Nevertheless, modern technology gives close control over these values, thus producing castings with fine detail, close tolerances and high strength.

Diagram illustrates the plunger mechanism which is submerged in molten metal. Modern machines are hydraulically operated and equipped with automatic cycling controls and safety devices.

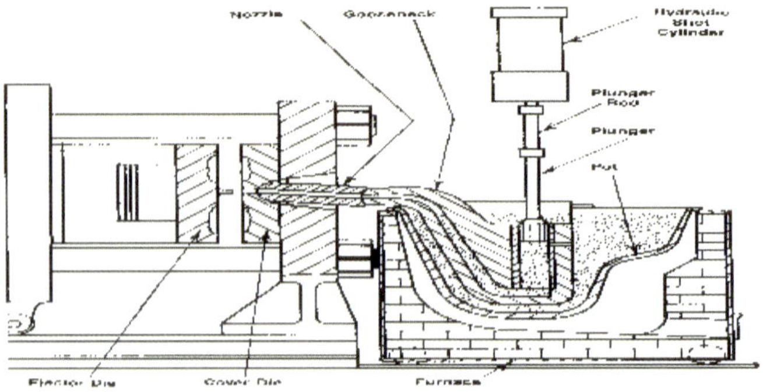


Figure: Hot chamber machine

## Cold Chamber Machines

Cold chamber machines differ from hot chamber machines primarily in one respect; the injection plunger and cylinder are not submerged in molten metal. The molten metal is poured into a "cold chamber" through a port or pouring slot by a hand or automatic ladle. A hydraulically operated plunger, advancing forward, seals the port forcing metal into the locked die at high pressures. Injection pressures range from 3,000 to over 10,000 psi for both aluminum and magnesium alloys, and from 6,000 to over 15,000 psi for copper-based alloys.

In a cold chamber machine, more molten metal is poured into the chamber than is needed to fill the die cavity. This helps sustain sufficient pressure to pack the cavity solidly with casting alloy. Excess metal is ejected along with the casting and is part of the complete shot.

Operation of a "cold chamber" machine is a little slower than a "hot chamber" machine because of the ladling operation. A cold chamber machine is used for high melting point casting alloys because plunger and cylinder assemblies are less subject to attack since they are not submerged in molten metal.

Diagram below illustrates die, cold chamber and horizontal ram or plunger (in charging position).

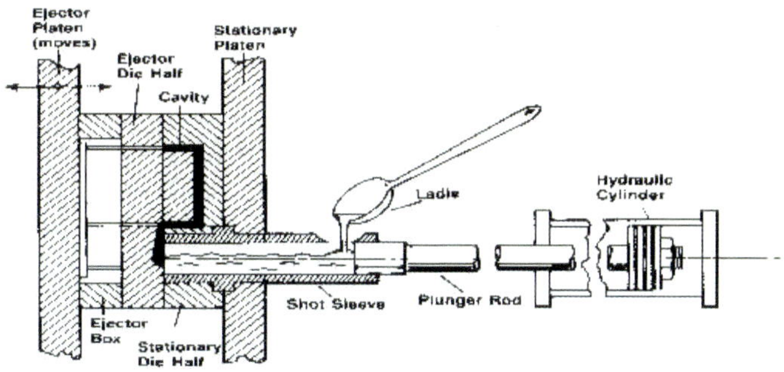


Figure: Cold chamber machine

## Casting Dies and their Construction

Die casting dies are made of alloy tool steels in at least two sections called fixed die half and ejector die half. The fixed die half is mounted on the side toward the molten metal injection system. The ejector die half, to which the die casting adheres, and from which it is ejected when the die is opened, is mounted on the moveable platen of the machine.

The fixed die half of the die is designed to contain the sprue hole through which molten metal enters the die. The ejector half usually contains the runners (passage ways) and gates (inlets) which route molten metal to the cavity (or cavities) of the die. The ejector half is also connected to an ejector box which houses the mechanism for ejecting the casting from the die. Ejection occurs when pins connected to the ejector plate move forward to force the casting from the cavity. This usually occurs as part of the opening stroke of the machine. Placement of ejector pins must be carefully arranged so force placed upon the casting during ejection will not cause deformation. Return pins attached to the ejector plate return this plate to its casting position as the die closes.

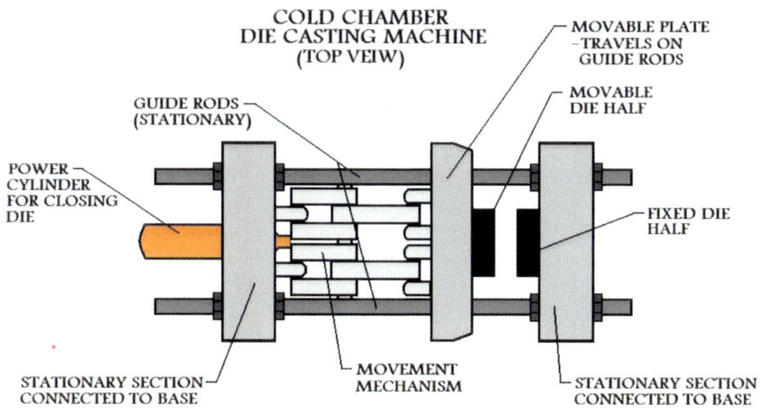


Fixed and moveable cores are often used in dies. If fixed, the core axis must be parallel to the direction of the die opening. If moveable, they are often attached to core slides. Should the side of a die casting design require a depression, the die can be made with one or more slides to obtain the desired result without affecting ejection of the casting from the die cavity. All moveable slides and cores must be carefully fitted, and have the ability to be securely locked into position during the casting cycle. Otherwise, molten metal could be forced into their slideways causing a disruption of operations. Although slides and cores add to the complexity and cost of die construction, they make it possible to produce die castings in a wide variety of configurations, and usually more economically than any other metalworking process.

## Die Terminology

Sprue holes are tapered with the small end located at the breaking point when the die is opened. A sprue pin, located in the ejector half, makes the sprue hollow and deflects metal entering the die into the runner system.

Runners are channels located at the parting line to route liquid metal from the sprue hole to the gate.

Gates are passages through which metal enters the die cavity. They have an important function in directing metal flow so that the cavity is correctly filled. Air is expelled through vents as molten metal enters the die cavity.

Guide pins assure proper alignment of die halves and correct register of cavities.

Side walls and cores are designed to have a slight taper or draft. The largest diameter or cross section of a cavity must be located at the parting line (unless slides are used) so the casting can be removed from the die. As molten metal in a die cavity solidifies, it shrinks away from walls onto core pins and other projections. The design of the die must permit withdrawal of core pins, and ejection without applying too much pressure which could cause deformation to the die casting.

Cores, fixed or moveable, as well as "loose pieces," must be positioned to facilitate removal either mechanically or by hand. Loose pieces used to form undercuts must be positioned by the operator. This requires extra labor, and usually slows the casting cycle.

Inserts can be cast integrally to provide special characteristics.

Most dies are cooled by water circulating through channels drilled for that purpose

## Automation in Die Casting

Automation and mechanization are terms often used interchangeably by the die casting industry. Automation of a manufacturing process is more sophisticated and complicated than the mechanization of various operations.

A die caster progresses towards automation by mechanizing various steps of the die casting process. For example:

- Die lubrication can be accomplished by installing fixed or reciprocating spray systems.
- An automatic ladling device can replace the hand ladle.
- Castings can be removed from the die by extractors or robots, or by "drop through" to a conveyor below the machine.
- Die casting machine operation can be integrated to cast, quench, trim and eject castings and return scrap to the furnace by conveyor.

By closing the loop, the die casting process can be automated, utilizing feed-back controls and microprocessors which automatically self compensate for process variables. As die casting plants move toward automation, noticeable changes occur. The operation tends to become safer. Production and quality improves considerably. Automation upgrades the labor force by requiring the development of new skills for machine set-up, programming, electronics, mechanics and maintenance.

### Quality in Die Casting

Quality in die casting is maintained through the use of process controls and feedback between the process control computer and the die casting machine.

Process controllers may utilize microprocessors to access transducers mounted on the die casting machine, to obtain velocity, position, hydraulic pressure and tie-bar strain data, etc. The microprocessor then adjusts the die casting machine operation through special valves, thus assuring consistent castings shot after shot. The process controller also collects machine performance data for statistical analysis in quality control.

## Continuous Casting

Continuous casting, also referred to as strand casting, is a process used in manufacturing industry to cast a continuous length of metal. Molten metal is cast through a mold, the casting takes the two dimensional profile of the mold but its length is indeterminate. The casting will keep traveling downward, its length increasing with time. New molten metal is constantly supplied to the mold, at exactly the correct rate, to keep up with the solidifying casting. Industrial manufacture of continuous castings is a very precisely calculated operation. Continuous casting can produce long strands from aluminum and copper, also the process has been developed for the production of steel.

### Process

Molten metal, from some nearby source, is poured into a tundish. A tundish is a container that is located above the mold, it holds the liquid metal for the casting. This particular casting operation uses the force of gravity to fill the mold and to help move along the continuous metal casting. The tundish is where the operation begins and is thus located high above ground level, as much as eighty or ninety feet. The continuous casting operation may require a lot of space.

It is the job of the tundish to keep the mold filled to the right level throughout the manufacturing operation. Since the metal casting is constantly moving through the mold, the tundish must always be supplying the mold with more molten metal to compensate.

The supplying of metal to the mold is not only going on throughout the entire manufacturing operation, it must be carried out with accuracy. A control system is employed to assist with this task. Basically the system can sense what the level of molten metal is, knows what the level should be, and can control the pouring of the metal from the tundish to ensure the smooth flow of the casting process. Although the tundish can typically hold several thousand pounds of metal, it too must be constantly supplied from the source of molten material.

The tundish also serves as the place where slag and impurities are removed from the melt. The high melting point and reactive nature, at high temperatures, has always made steel a difficult material to cast. When a manufacturing operation is continuously casting steel, the reactivity of the molten steel to the environment needs to be controlled. For this purpose, the mold entrance may be filled with an inert gas such as argon. The inert gas will push away any other gases, such as oxygen, that may react with the metal. There is no need to worry about the inert gas reacting with a molten metal melt, since inert gases do not react with anything at all.

The metal casting moves quickly through the mold, in the continuous manufacture of the metal part. The casting does not have time to solidify completely in the mold. As can be remembered from our discussion on solidification, a metal casting will first solidify from the mold wall, or outside of the casting, then solidification will progress inward. The mold in the continuous casting process is water cooled, this helps speed up the solidification of the metal casting. As stated earlier, the continuous casting does not completely harden in the mold. It does, however, spend enough time in the water cooled mold to develop a protective solidified skin of an adequate thickness on the outside.

The long metal strand is moved along at a constant rate, by way of rollers. The rollers help guide the strand and assist in the smooth flow of the metal casting out of the mold and along its given path. A group of special rollers may be used to bend the strand to a 90 degree angle. Then another set will be used to straighten it, once it is at that angle. Commonly used in manufacturing industry, this process will change the direction of flow of the metal strand from vertical to horizontal.

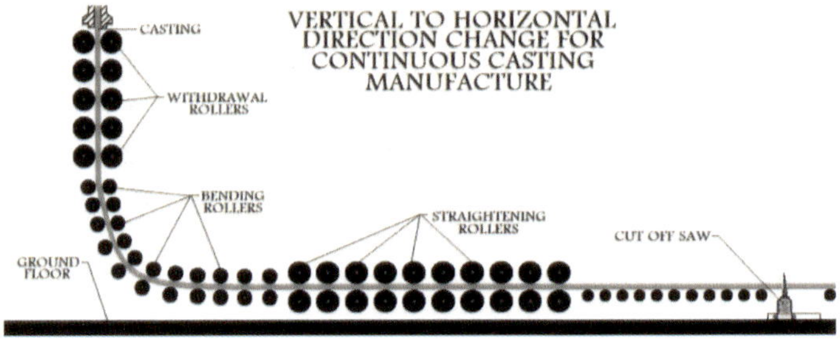


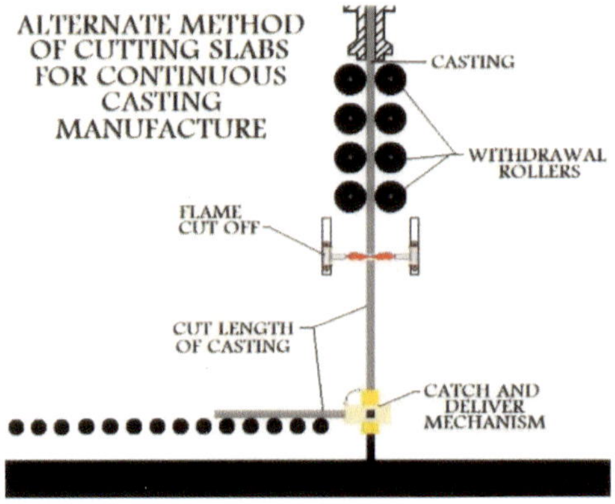


The continuous casting can now travel horizontally as far as necessary. The cutting device, in manufacturing industry, is typically a torch or a saw. Since the metal casting does not stop moving, the cutting device must move with the metal casting, at the same speed, as it does its cutting. There is another commonly used setup for cutting lengths of metal casting strand from a continuous casting operation. This particular manufacturing setup eliminates the need for bending and straightening rollers. It does, however, limit the length of metal casting strand that may be produced, based in a large part on the height of the casting floor where the mold is located.

There needs to be an initial setup for a continuous casting operation, since you can not just pour molten metal through an empty system to start off the process. To begin continuous casting manufacture, a starter bar is placed at the bottom of the mold. Molten material for the metal casting is poured into the mold and solidifies to the bar. The bar gives the rollers something to grab onto initially. The rollers pull the bar, which pulls along the continuous casting.

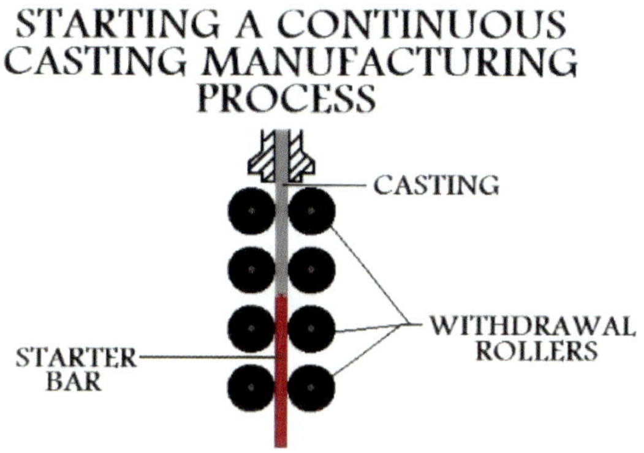


In the manufacture of a product, often two or more different kinds of operations may need to be performed. Such as a metal casting operation followed by a metal forming operation. In modern commercial industry, the continuous casting process can be integrated with metal rolling. Do not confuse the rolling operation with the rolls used to guide the casting. The rolling operation is a forming process and it will change the metal it processes. Rolling of the metal strand, is the second manufacturing process and it must be performed after the casting operation. Continuous casting is very convenient in that the rolling mill can be fed directly from the continuously cast metal casting strand. The metal strand can be rolled directly into a given cross sectional shape such as an I beam. The rate of the rolling operation is synchronized with the speed that the continuous metal casting is produced and thus the two operations are combined as one.

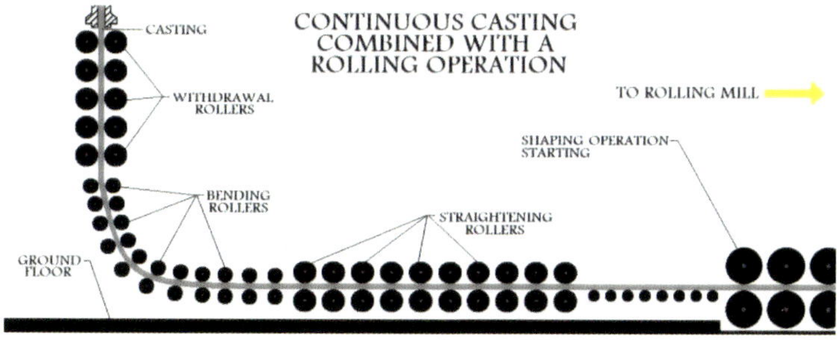


## Properties and Considerations of Manufacturing by Continuous Casting

- Continuous casting manufacture is different from other metal casting processes, particularly in the timing of the process. In other casting operations, the different steps to the process such as the ladling of metal, pouring, solidification, and casting removal all take place one at a time in a sequential order. In continuous casting manufacture, these steps are all occurring constantly and at the same time.

- This process is used in commercial manufacture as a replacement to the traditional process of casting ingots.
- Piping, a common problem in ingot manufacture, is eliminated with the continuous casting process.
- Structural and chemical variations in the metal of the casting, often present in ingots, have been eliminated. When manufacturing with the continuous metal casting process, the casting's material will possess uniform properties.
- When employing continuous metal casting manufacture, the castings will solidify at 10 times the rate that a casting solidifies during ingot production.
- With less loss of material, cost reduction, higher productivity rate, and superior quality of castings, continuous casting manufacture is often the choice over ingot production.
- A continuous casting manufacturing process will take considerable resources and planning to initiate, it will be employed in only very serious industrial operations.

## Centrifugal Casting

Centrifugal casting, sometimes called rotocasting, is a metal casting process that uses centrifugal force to form cylindrical parts. This differs from most metal casting processes, which use gravity or pressure to fill the mold. In centrifugal casting, a permanent mold made from steel, cast iron, or graphite is typically used. However, the use of expendable sand molds is also possible. The casting process is usually performed on a horizontal centrifugal casting machine (vertical machines are also available) and includes the following steps:

1. Mold preparation - The walls of a cylindrical mold are first coated with a refractory ceramic coating, which involves a few steps (application, rotation, drying, and baking). Once prepared and secured, the mold is rotated about its axis at high speeds (300-3000 RPM), typically around 1000 RPM.
2. Pouring - Molten metal is poured directly into the rotating mold, without the use of runners or a gating system. The centrifugal force drives the material towards the mold walls as the mold fills.
3. Cooling - With all of the molten metal in the mold, the mold remains spinning as the metal cools. Cooling begins quickly at the mold walls and proceeds inwards.
4. Casting removal - After the casting has cooled and solidified, the rotation is stopped and the casting can be removed.
5. Finishing - While the centrifugal force drives the dense metal to the mold walls, any less dense impurities or bubbles flow to the inner surface of the casting. As a result, secondary processes such as machining, grinding, or sand-blasting, are required to clean and smooth the inner diameter of the part.

Centrifugal casting is used to produce axi-symmetric parts, such as cylinders or disks, which are typically hollow. Due to the high centrifugal forces, these parts have a very fine grain on the outer surface and possess mechanical properties approximately 30% greater than parts formed with static casting methods. These parts may be cast from ferrous metals such as low alloy steel, stainless steel, and iron, or from non-ferrous alloys such as aluminum, bronze, copper, magnesium, and nickel. Centrifugal casting is performed in wide variety of industries, including aerospace, industrial, marine, and power transmission. Typical parts include bearings, bushings, coils, cylinder liners, nozzles, pipes/tubes, pressure vessels, pulleys, rings, and wheels.

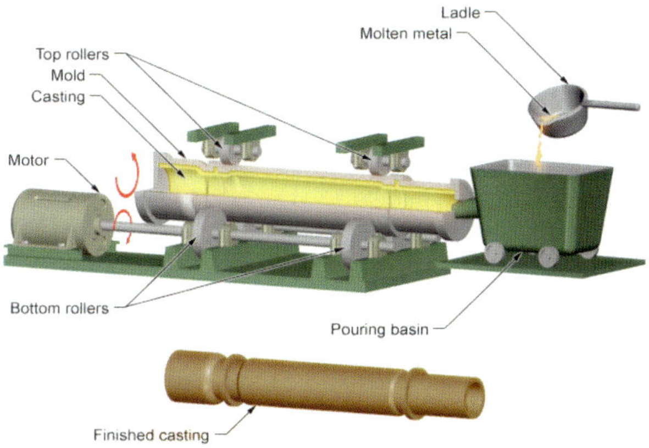


Centrifugal Casting

## Types

- True Centrifugal Casting

True centrifugal casting is sometime, known as centrifugal casting is a process of making symmetrical round hollow sections. This process uses no cores and the symmetrical hollow section is created by pure centrifugal action. In this process, the mold rotates about horizontal or vertical axis. Mostly the mold is rotated about horizontal axis and the molten metal introduce from an external source. The centrifugal force acts on the molten metal, which forces it at the outer wall of mold. The mold rotates until the whole casting solidifies. The slag particles are lighter than metal thus separated at the central part of the casting and removed by machining or other suitable process. This process used to make hollow pipes, tubes, and hollow bushes etc., which are axi-symmetrical with a concentric hole.

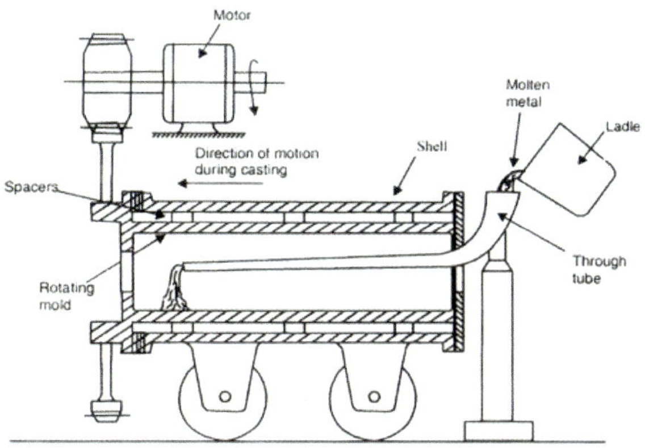

- Semi Centrifugal Casting

This process is used to cast large size axi symmetrical object. In this process mold is placed horizontally and rotated along the vertical axis. A core is inserted at the center which is used to cast hollow section. When the mold rotates, the outer portion of the mold fill by purely centrifugal action and as the liquid metal approaches toward the center, the centrifugal component decreases and gravity component increase. Thus a core is inserted at center to make hollow cavity at the center without centrifugal force. In this process centrifugal force is used for uniform filling of axi symmetrical parts. Gear blanks, flywheel etc. are made by this process.

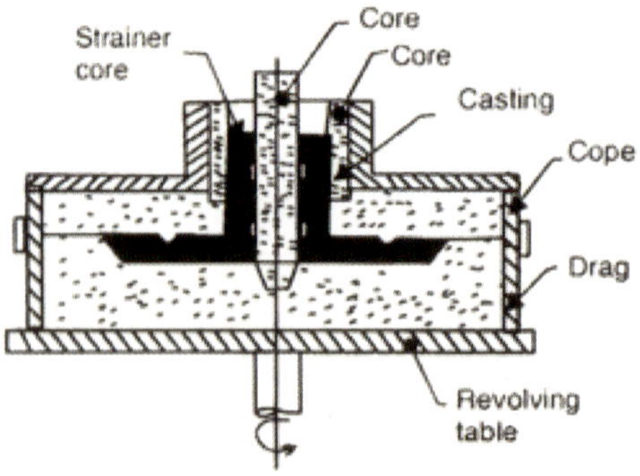


- Centrifuging

In this process there are several mold cavities connected with a central sprue with radial gates. This process uses higher metal pressure during solidification. It is used to cast shapes which are not axi symmetrical. This is only suitable for small objects.

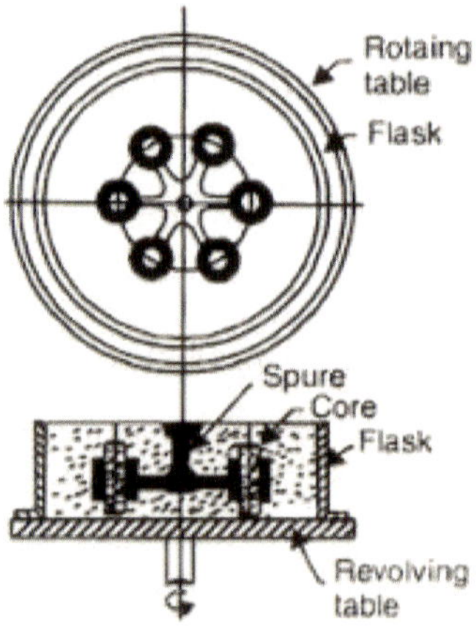


## Application

- It is widely used in aircraft industries to cast rings, flanges and compressor casting.
- It is used for cast Steam turbine bearing shell.
- Roller for steel rolling mill is another example of centrifugal casting.
- It is used in automobile industries to cast gear blank, cylindrical liners, piston rings etc.

- It is used to cast bearings.
- This process used to cast switch gear components used in electronic industries.

## Advantages

- It provides dense metal and high mechanical properties.
- Unidirectional solidification can obtain up to a certain thickness.
- It can use for mass production.
- No cores are required for cast hollow shapes like tubes etc.
- Gating system and runner are totally eliminated.
- All the impurity like oxide or other slag particles, segregated at center from where it can easily remove.
- It required lower pouring temperature thus save energy.
- Lower casting defects due to uniform solidification.

## Disadvantages

- Limited design can be cast. It can cast only symmetrical shapes.
- High equipment or setup cost.
- It is not suitable for every metal.
- Higher maintenance required.
- High skill operator required.

In this casting process, solidification time and temperature distribution is difficult to determine.

Centrifugal casting of lead-free pewter offers jewelry designers a lower cost alternative to sterling silver and bronzes. It is especially a good fit for large bulky pieces where metal content and weight would detrimentally impact cost.

Centrifugal casting systems clamp a two piece mold between two platens holding the mold firmly and in parallel to each other to ensure that the edges of the mold are sealed and the integrity of the cavity sound. The mold will then spin while molten pewter is then poured into the feed tube. The metal is pulled by the centrifugal force to the extremes of the mold and fills from the outer edge towards the feed tube. Solidification occurs while molten metal is still filling. The transition from metal to solid and the resulting shrink is backfilled by additional liquid metal. The result is a cast piece that has very little porosity and a smooth surface.

Lead-free pewter is the safest inexpensive pewter metal for jewelry applications. Keeping lead out of metals that will come in contact with the skin eliminates the chance that someone will pick up lead in their body as a result. This is critical for women's jewelry where small children may put them in the mouth when being held.

## Spin Casting

Spin Casting is also known as centrifugal rubber mold casting (CRMC).

The main features of spin casting process is the use of rubber mold and employing centrifugal force for casting. During the solidification process the mold spins on it central axis till the metal solidifies in the mold. Spin casting is a widely used non-expendable process, for casting products that have excellent details and surface finish.

## Process

A spin casting begins with a mold layout and selection of pattern material. The layout of a spin casting mold will usually consist of multiple parts that are placed symmetrically around the center hub. A mold may be designed to create many copies of the same part or many different parts. This layout will determine the number of FDM patterns required.

Of all the Fortus materials, PC and PPSF are the best suited for pattern making. Due to the temperature and pressure applied to the mold during vulcanization, patterns constructed in other Fortus materials may warp and distort. In general, PPSF will produce the most accurate spin cast parts because of its high heat defl ection temperature and mechanical strength.

However, it also requires more time and labor for support removal and pattern preparation. PC is more easily finished, but it is slightly less stable in the vulcanizing process, which may translate to larger dimensional deviations.

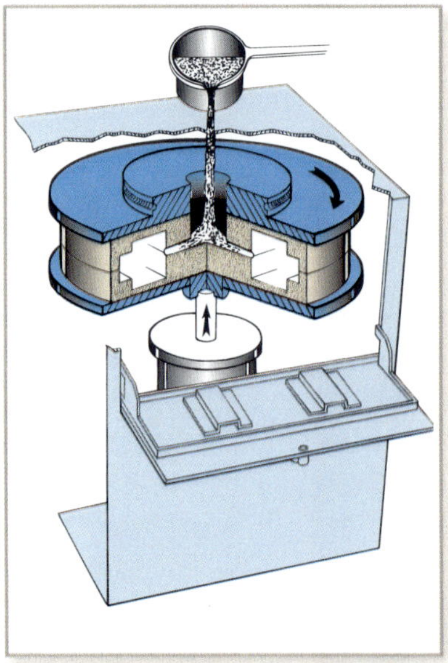

Figure: Metal, plastic or wax is poured into a rotating mold. Centrifugal force pushes the material into the mold cavities.

## Pattern Making

FDM patterns can be made from a casting's CAD data with no need for modifi cation. Since the mold is pliable, draft angles do not need to be added to the design and small undercuts do not need to be removed. Optionally, shrinkage compensation can be added to the CAD data prior to exporting an STL fi le, but this can also be done within Insight.

The shrinkage compensation will vary with the rubber used for the mold and the material that is cast. Refer to supplier information and calculate the net shrinkage for the mold and castings.

Scale the STL files by this shrinkage amount. In Insight, orient the patterns for the best surface quality and detail, and then select the solid build style. Any patterns constructed with sparse fill will be subject to collapse when exposed to the pressure of the vulcanizer.

After the build is completed, remove the support structures and finish the patterns to the desired quality level. Since the rubber molds will pick up very small details, it is import to smooth all surfaces to the quality level needed in the cast parts. To achieve the desired finish, use a combination of DCM (methylene chloride) dipping (PC only), sanding, filling and priming.

## Mold Making

The spin casting mold begins as pre-formed, uncured rubber disks that have a consistency similar to modeling clay. The type of rubber—organic or silicone—is selected based on the material to be cast, type of part and desired production quantities.

Figure: Patterns, center plug and locknuts are placed on the uncured rubber disk.

The rubber disks are stacked to the desired thickness for the core side of the mold. The patterns are then arranged in a balanced, symmetrical pattern to ensure even material distribution. Each pattern is then aligned such that material can be easily pulled through the cavity.

Typically, this orientation has the longest side of the part along a radial line from the hub.

Once placed and aligned, the patterns are then embedded in the rubber to define the parting line for the cast part. For flat bottomed parts, the patterns are laid on top of the rubber. For all others, a shallow pocket is cut into the rubber. The pattern is then set into the pocket, and the excess rubber is shaped around it to establish the parting line.

Next, insert a center plug into the middle of the rubber disc to create the sprue. Then arrange lock-nuts or pins on the perimeter to ensure proper alignment of the two mold halves when assembled for casting. Optionally, preforms for the runner system may also be placed into the mold. The core side of the mold is now complete. Place the core side in a circular mold frame, and spray the surface with mold release. To complete the mold, stack additional uncured rubber discs on top of the core side of the mold. This will be the cavity side of the mold.

## Mold Vulcanizing

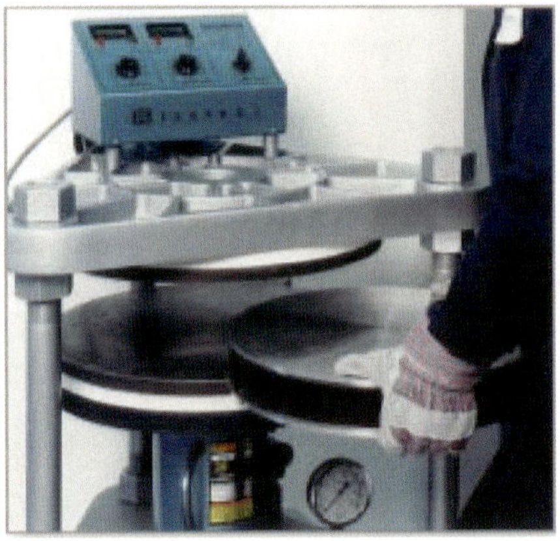

Figure: The mold is loaded into a vulcanizer that cures the rubber.

The vulcanizer consists of two heated platens mounted on a hydraulic press. The heat and pressure of the vulcanizer cause the uncured rubber to flow around the patterns and fill the voids. As the exposure to elevated temperatures continues, the rubber begins to cure, which causes it to become firm yet flexible.

The uncured rubber mold containing the patterns is placed into the vulcanizer, which is preheated to 315°F (157°C). The pressure is then slowly raised to approximately 1,000 psi (6.9 MPa) to squeeze the halves of the mold. The pressure and temperature, which vary by type of rubber, are maintained for one to two hours.

Figure: Gates, runners and vents are cut into the rubber mold with a sharp knife or scalpel.

When vulcanizing is complete, the mold is removed. After a short cooling period, the mold frame is taken off, and the two halves are separated. The patterns, and any metal preforms, are then extracted from the mold. Gates, runners and vents are now cut into the cured rubber with a sharp knife or scalpel. Typically, each is a V-shaped channel. The gates and runners feed casting material to the part cavity from the central hub. The vent allows air in the cavity to escape so that back pressure does not cause a partial fi ll of the mold cavity. The mold is now ready for spin casting.

## Casting

Figure : The mold is loaded into the spin casting machine.

Apply mold release to both sides of the rubber mold, close the mold and place it in the spin casting machine. Prior to starting the spin casting machine, prepare the casting material. If casting metal, melt the material in a gas or electric furnace. Bring the molten alloy to the ideal casting temperature. If the metal is too cold, it will freeze off before fi lling the mold, and if too hot, it will degrade the mold prematurely. When casting foundry wax, melt the material in a suitable melting tank or pot. For thermoset materials, combine the two parts of the material kit and stir thoroughly.

To prepare the spin caster, select the rotational speed, clamping pressure and cycle time. Each variable will be dependent on the material that is cast. For example, metals will have a cycle time of less than one minute, while plastics will have a duration of 5 to 10 minutes.

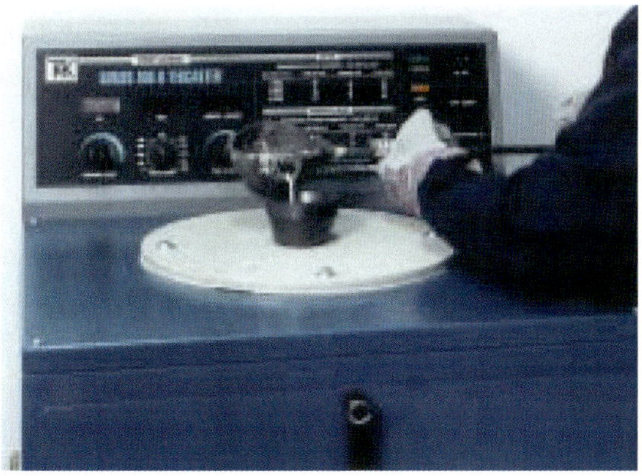

Figure: While spinning, molten metal is poured into the mold.

Start the spin caster, and as the mold is spinning, pour the casting material into the funnel at the top of the machine. When the cycle is complete, remove the mold from the spin caster.

## Finishing

Separate the two halve of the rubber mold to expose the castings. To extract them, flex the rubber or gently pry the casting from its cavity. If any material remains in the gate, runner or vent channels, remove it prior to reusing the mold. Finish the casting by snapping the gates off of the part and grinding or sanding the remainder. The castings are now ready for painting, plating or use.

Applications: Spin casting is applicable for a wide variety of industrial, decorative, novel items, fashion accessories, etc. Plaster and metals with low melting poring and viscosities like pewter, zinc alloys, etc can be spin casted. Some of the products that are spin casted are inserts such as those used in fishing lure manufacturing, fishing tackle components, belt buckles, pins, emblems, medallions, trophies, assorted souvenirs etc. In addition, effective models of fragile rapid prototype models can be quickly made.

## Advantage:

- Cast parts may be finished or painted as desires
- A quick casting process
- In a mold around 50 small to medium castings can be made in an hour
- Parts of any complexity can be handled
- High integrity, close tolerance parts with excellent surface finish are produced
- Low start up cost hence conducive for hobbyist and small business ventures.

## References


- What-is-casting-working-process-advantages-disadvantages-terminology-and-application: mech4study.com, Retrieved 26 July 2018
- Manufacturing-technology, what-is-casting-39480: Retrieved 21 June 2018
- Casting-lost-foam-casting: industrialmetalcastings.com, Retrieved 11 April 2018
- Advantages-of-permanent-mold-castings-and-products: buntingbearings.com, Retrieved 26 May 2018
- Permanent-mold-process-limitation: chinasavvy.com, Retrieved 20 March 2018
- Die-casting-faqs/what-is-die-casting, die-casting-101: paceind.com, Retrieved 10 March 2018
- Centrifugal-casting-of-lead-free-pewter-jewelry: gsjewelrymfg.com, Retrieved 09 June 2018

# Molding Techniques

Molding is a manufacturing process concerned with the shaping of a liquid or a raw material by using a rigid frame called a matrix or mold. All the diverse techniques of molding such as blow molding, transfer molding, injection molding, compression molding, rotational molding and thermoforming, have been carefully analyzed in this chapter.

A mold is formed into the geometric shape of a desired part. Molten metal is then poured into the mold, the mold holds this material in shape as it solidifies. A metal casting is created. Although this seems rather simple, the manufacturing process of metal casting is both a science and an art. Let's begin our study of metal casting with the mold. First, molds can be classified as either open or closed. An open mold is a container, like a cup, that has only the shape of the desired part. The molten material is poured directly into the mold cavity which is exposed to the open environment.

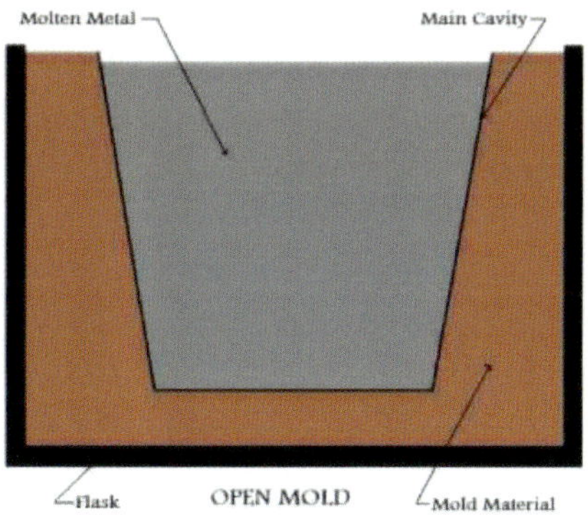


This type of mold is rarely used in manufacturing production, particularly for metal castings of any level of quality. The other type of mold is a closed mold, it contains a delivery system for the molten material to reach the mold cavity, where the part will harden within the mold. A very simple closed mold is shown in figure. The closed mold is, by far, more important in manufacturing metal casting operations.

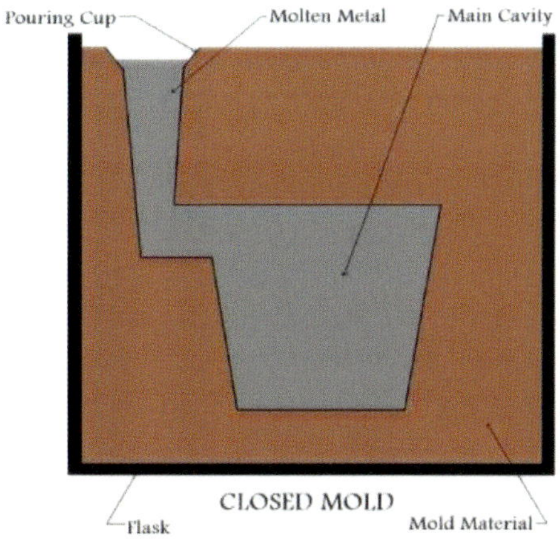

There are many different metal casting processes used in the manufacture of parts. Two main branches of methods can be distinguished by the basic nature of the mold they employ. There is expendable mold casting and permanent mold casting. As the name implies, expendable molds are used for only one metal casting while permanent molds are used for many. When considering manufacturing processes, there are advantages and disadvantages to both.

## Expendable Mold

- Can produce one metal casting only
- Made of sand, plaster, or other similar material
- Binders used to help material hold its form
- Mold that metal solidifies in must be destroyed to remove casting
- More intricate geometries are possible for casting

## Permanent Mold

- Can manufacture many metal castings
- Usually made of metal or sometimes a refractory ceramic
- Mold has sections that can open or close, permitting removal of the casting
- Need to open mold limits part shapes

## Patterns

Expendable molds require some sort of pattern. The interior cavities of the mold, in which the molten metal will solidify, are formed by the impression of this pattern. Pattern design is crucial to success in manufacture by expendable mold metal casting. The pattern is a geometric replica of the metal casting to be produced. It is made slightly oversize to compensate for the shrinkage that will occur in the metal during the casting's solidification, and whatever amount of material that will be machined off the cast part afterwards. Although machining will add an extra process to the manufacture of a part, machining can improve surface finish and part dimensions considerably. Also, increasing the machine finish allowance will help compensate for unknown variables in shrinkage, and reduce trouble from areas of the metal casting that may have been originally too thin or intricate.

## Pattern Material

The material from which the pattern is made is dependent upon the type of mold and metal casting process, the casting's geometry and size, the dimensional accuracy required, and the number of metal castings to be manufactured using the pattern. Patterns can be made from wood, like pine (softwood), or mahogany (hardwood), various plastics, or metal, like aluminum, cast iron, or steel. In most manufacturing operations, patterns will be coated with a parting agent to ease their removal from the mold.

## Cores

For metal castings with internal geometry cores are used. A core is a replica, (actually an inverse), of the internal features of the part to be cast. Like a pattern, the size of the core is designed to accommodate for shrinkage during the metal casting operation. Unlike a pattern, a core remains in the mold while the metal is being poured. Hence, a core is usually made of a similar material as the mold. Once the metal casting has hardened, the core is broken up and removed much like the mold. Depending upon the location and geometry of the core within the casting, it may require that it is supported during the operation to prevent it from moving or shifting. Structural supports that hold the core in place are called chaplets. The chaplets are made of a material with a higher melting temperature than the casting's material, and become assimilated into the part when it hardens. Note that when manufacturing a metal casting with a permanent mold process, the core will be a part of the mold itself.

## Mold

A typical mold is shown in figure.

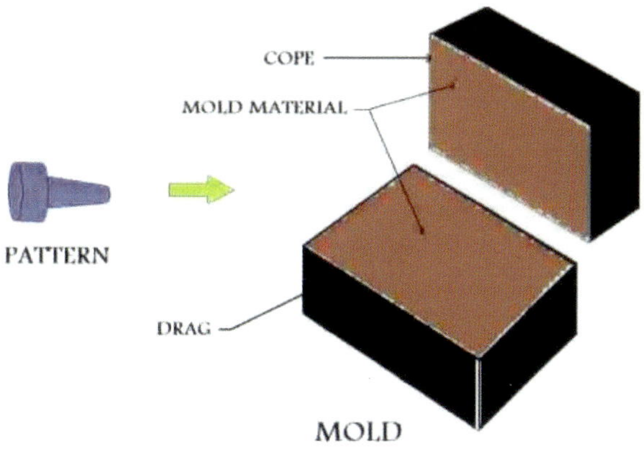


When manufacturing by metal casting, consideration of the mold is essential. The pattern is placed in the mold and the mold material is packed around it. The mold contains two parts, the drag (bottom), and the cope (top). The parting line between the cope and drag allows for the mold to be opened and the pattern to be removed once the impression has been made.

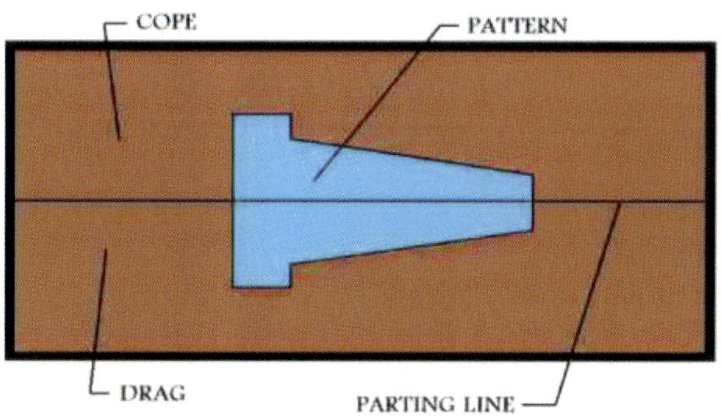


The core is placed in the metal casting after the removal of the pattern. Figure shows the pattern impression with the core in place.

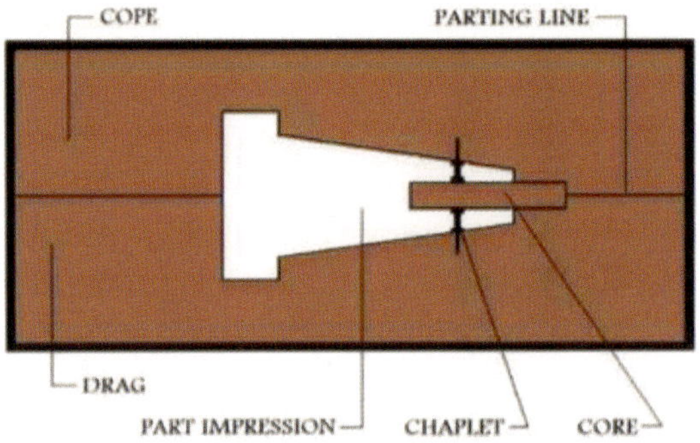


Now the impression in the mold contains all the geometry of the part to be cast. This metal casting setup, however, is not complete. In order for this mold to be functional to manufacture a casting, in addition to the impression of the part, the mold cavity will also need to include a gating system. Sometimes the gating system will be cut by hand or in more adept manufacturing procedures, the gating system will be incorporated into the pattern along with the part. Basically, a gating system functions during the metal casting operation to facilitate the flow of the molten material into the mold cavity.

## Elements of a Gating System

This is where the molten metal employed to manufacture the part enters the mold. The pouring basin should have a projection with a radius around it to reduce turbulence.

From the pouring basin, the molten metal for the casting travels through the down sprue. This should be tapered so its cross-section is reduced as it goes downward.

The down sprue ends at the sprue base. It is here that the casting's inner cavity begins.

Once at the sprue base, the molten material must pass through the ingate in order to enter the inner area of the mold. The ingate is very important for flow regulation during the metal casting operation.

Runners are passages that distribute the liquid metal to the different areas inside the mold.

The impression of the actual part to be cast is often referred to as the main cavity.

Vents help to assist in the escape of gases that are expelled from the molten metal during the solidification phase of the metal casting process.

Risers are reservoirs of molten material. They feed this material to sections of the mold to compensate for shrinkage as the casting solidifies. There are different classifications for risers,

Top Risers: Risers that feed the metal casting from the top.

Side Risers: Risers that feed the metal casting from the side.

Blind Risers: Risers that are completely contained within the mold.

Open Risers: Risers that are open at the top to the outside environment.

Figure shows a mold with all its features, ready for metal casting.

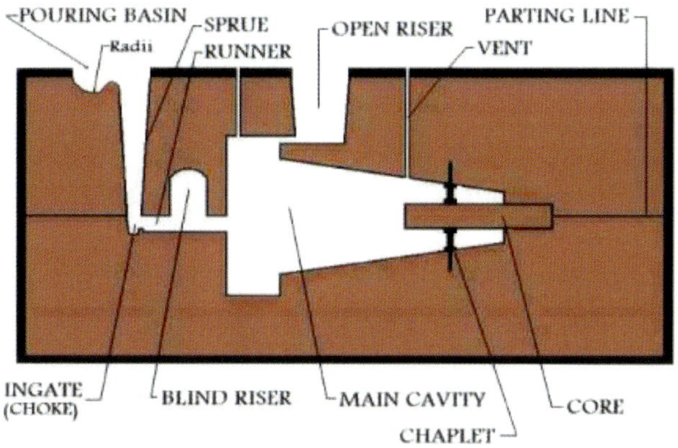


## Types of Molding Processes

Different types of molding processes form plastic into the desired shape based on the plastic's intended use. Plastic manufacturing relies on different types of molding in a variety of shapes. Plastic is a synthetic material that is soft or semi-liquid when hot. The soft plastic is placed in molds, and then the plastic cools, or sets. After setting, the plastic is in the desired shape and is removed from whatever types of molding were used to create it. Thermoplastics can be melted down and reformed if necessary, however thermoset plastics cannot be reheated.

## Plastic Molding using Casting

Plastic molding using casting is the simplest method as plastic manufacturing as it requires the least amount of complex technology. Plastic is simply heated so it turns into a fluid, and then transferred into a mold. The plastic is left to cool and the mold is removed. This process can be used for intricate shapes and is performed under low pressure.

## Injection Molding of Plastic

Injection molding of plastic creates high-quality three-dimensional objects that can be commercially reproduced. The injection molding process begins by melting plastic in a hopper. The melted, liquid plastic is injected into a tightly closed, chilled mold. The plastic quickly takes the shape of the surrounding mold. Once it has completely set, the mold is opened to release the plastic object. The mold can generally be used many times before needing to be replaced. Plastic items such as yogurt cups, butter tubs, plastic toys and bottle caps use the injection molding process.

## Blow Molding Plastic Manufacturing

Blow molding is a process used for making hollow objects such as piping or milk bottles. In the blow molding plastic manufacturing process, plastic is heated until molten. The liquid, molten plastic is injected into a cold mold. The mold has a tube set within it, which has a particular shape when inflated. While the plastic is molten, air is blown into the tube and the plastic is formed around the tubing. The plastic is left to cool and removed from the mold.

## Compression Molding of Plastic

Compression molding of plastic is the most labor-intensive type of molding process. Since compression molding is more complicated, it is typically only used for large-scale production purposes rather than mass production. For example, boat hulls and car tires are made using the compression molding method. Molten plastic is poured into a mold. Then a second mold is pressed into it. This squeezes the plastic into the desired shape before the plastic is left to cool and removed from the mold.

## Rotational Molding of Plastic

Toys, shipping drums, storage tanks and consumer furniture made of plastic are all made using rotational molding. In this method, liquid plastic forms each object as it is added to the mold from the inside. Two mechanical arms hold the mold in place. The arms constantly rotate the mold at the same level, while molten plastic is placed inside. As the mold turns, the plastic coats the inside of the mold to create a new hollow, plastic object.

# Blow Molding

Blow molding is the process of forming a molten tube (referred to as the parison or preform) of thermoplastic material (polymer or resin) and placing the parison or preform within a mold cavity and inflating the tube with compressed air, to take the shape of the cavity and cool the part before removing from the mold.

Any hollow thermoplastic part can be blow molded.

Parts are not just limited to bottles, where there is one opening and it is usually smaller in diameter or size than the overall body dimensions. These are some of the most common shapes used in consumer packaging, however there are other typical types of blow molded parts, including, but not limited to:

- Industrial bulk containers
- Lawn, garden and household items
- Medical supplies and parts, toys
- Building industry products
- Automotive-under the hood parts
- Appliance components

The basic concept of blow molding comes from the glassblowing process. As early as the 1880's there was a patented method of extruding a Celluloid polymer into a parison and blow molding a part. Early methods and materials to blow plastics into shapes were very crude and not suitable for mass production. It wasn't until the late 1930's that the first commercial machines and products were developed for the manufacturing of blow molded bottles. The major event that triggered and fueled the early years of commercial blow molding was the development of low and high density polyethylene suitable for blow molding and consumer packages. From this, there was an explosion of blow molded products and equipment in both Europe and North America.

## Blow Molding Manufacturing Processes

There are three main types of blow molding:

- Extrusion blow molding
- Injection blow molding
- Injection stretch blow molding

The main differences among them are the method of forming the parison; either by extrusion or injection molding, the size of the parison and the method of movement between the parison and blow molds; either stationary, shuttling, linear or rotary.

In extrusion blow molding-(EBM) the polymer is melted and the solid extruded melt is extruded through a die to form a hollow tube or parison. Two halves of a cooled mold is then closed around the parison, pressurized air is introduced through a pin or needle, inflating it into the shape of mold, thus producing a hollow part. After the hot plastic has cooled sufficiently, the mold is opened and the part is removed.

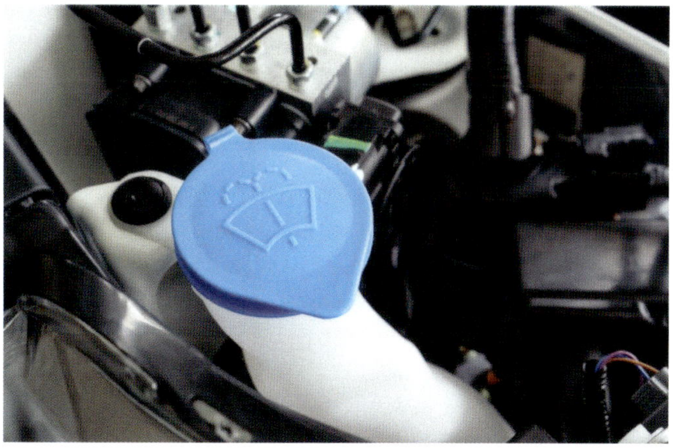

In EBM there are two basic methods of extrusion, continuous and intermittent. In continuous, the parison is extruded continuously and the mold moves to and away from the parison. In Intermittent, plastic is accumulated by the extruder in a chamber, then forces through the die to form the parison. The molds are typically stationary under or around the extruder.

Examples of the Continuous Process are Continuous Extrusion Shuttle machines and Rotary

Wheel machines. Intermittent extrusion machines can be Reciprocating Screw or Accumulator Head. Various factors are considered when selecting between the processes and the size or models available.

Examples of parts made by the EBM process include many hollow products, such as bottles, industrial parts, toys, automotive, appliance components and industrial packaging.

With respect to the injection blow systems – (IBS) process, the polymer is injection molded onto a core within a cavity to form a hollow tube called a preform. The preforms rotate on the core rod to the blow mold or molds at the blowing station to be inflated and cooled. This process is typically used to make small bottles, usually 16oz/500ml or less at very high outputs. The process is divided into three steps: injection, blowing and ejection, all done in an integrated machine. Parts come out with accurate finished dimensions and capable of holding tight tolerances—with no extra material in the formation it is highly efficient.

Examples of IBS parts are pharmaceutical bottles, medical parts, and cosmetic and other consumer product packages.

Injection stretch blow molding- (ISBM) process is similar to the IBS process described above, in that the preform is injection molded. The molded preform is then presented to the blow mold in a conditioned state, but before final blowing of the shape, the preform is stretched in length as well as radially. The typical polymers used are PET and PP that have physical characteristics that are enhanced by the stretching part of the process. This stretching gives the final part improved strength and barrier properties at much lighter weights and better wall thicknesses than IBS or EBM—but, not without some limits such as handled containers, etc.. ISBM can be divided into the One Step and Two Step process.

In the One Step process both preform manufacture and bottle blowing are performed in the same machine. This can be done in 3 or 4 station machines, (Injection, Conditioning, Blowing and Ejection). This process and related equipment can handle small to high volumes of various shape and size bottles.

In the Two Step process the plastic is first molded into the preform using an injection molding machine separate from the blow molder. These are produced with the necks of the bottles, including threads on the open end of the closed end hollow preform. These preforms are cooled, stored, and fed later into a re-heat stretch blow molding machine. In the Two Step Reheat Blow process, the preforms are heated (typically using infrared heaters) above their glass transition temperature, then stretched and blown using high-pressure air in the blow molds.

The Two Step process is more suited to very high volumes of containers, 1 liter and under, with very conservative use of resin providing great strength, gas barrier and other features.

## Materials Suitable for Blow Molding

Examples of Blow Molding Polymers in single layer or multiple layer structures or combinations include such materials as: (i) High Density Polyethylene, (ii) Low density Polyethylene, (iii) Polypropylene, (iv) Co-polyester, (v) PET, (vi) PVC, (vii) Nylon, (viii) EVOH, (ix) EVA, (x TPE, (xi) COP and COC, (xii) Polycarbonate, (xiii) Polystyrene, (ix) ABS, etc.

# Transfer Molding

The process of transfer molding is one in which the material which is to be used for carrying out the molding process is first weighed and then poured into the container used for the process of molding.

## Process of Transfer Molding

In the transfer molding, polymer charge is transferred from the transfer pot to the mold. The mold is cooled and molded part is ejected. The schematic of transfer molding process is shown in figure.

In this process, the required amount of polymer charge is weighted and inserted into the transfer pot before the molding process. The transfer pot is heated by the heating element above the melting temperature of the polymer charge. The liquid charge is gravity filled through the sprue to the mold cavity. A "piston and cylinder" arrangement is built in the transfer pot so that the resin is squirted into the mold cavity through a sprue. The plunger is also preheated in the transfer pot. The plunger is used to push the liquid polymer charge from the transfer pot into the mold cavity under pressure. The mold cavity remains closed as the polymer charge is inserted.

The mold cavity is held closed until the resin gets cured. The mold cavity is opened and the molded part can be removed once it has hardened with the help of ejector pin. The sprue and gate attached to the molded part have to be trimmed after the process has been completed.

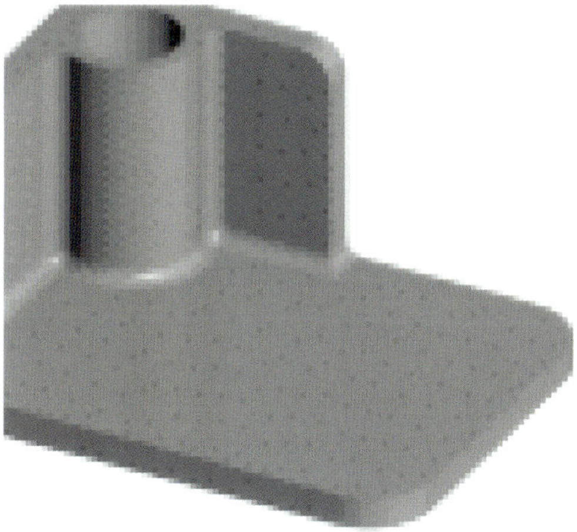

Figure: Transfer molding process

This is used for mass production. It has short production cycle and smaller tolerances and more intricate parts can be achieved. It produces more waste material; therefore it is the more expensive process. The mold cavity can be made from metals such as aluminum or steel for larger production.

## Process Parameters

- Heating time
- Melting temperature of the charge

- Applied pressure
- Cooling time

### Materials used

Generally, thermoset plastics (such as epoxy, polyester, phenol-formaldehyde, vinyl ester, silicone) are processed by transfer molding process, but certain thermoplastic materials can also be processed.

### Applications

This process is widely used to encapsulate items such as integrated circuits, plugs, connectors, pins, coils, and studs. It is suitable for molding with ceramic or metallic inserts which are placed in the mold cavity. When the heated polymer fills the mold it forms bonding with the insert surface. Transfer molding is also used for manufacturing radio and television cabinets and car body shells.

### Advantages

- Fast setup time and lower setup costs
- Low maintenance cost
- Plastic parts with metal inserts can be made
- Design flexibility
- Dimensionally stable
- Uniform thickness of parts
- Large production rate

### Disadvantage

- Wastage of material
- Production rate lower than injection molding
- Air can be trapped in the mold

## Injection Moulding

Injection molding is the most commonly used manufacturing process for the fabrication of plastic parts. A wide variety of products are manufactured using injection molding, which vary greatly in their size, complexity, and application. The injection molding process requires the use of an injection molding machine, raw plastic material, and a mold. The plastic is melted in the injection molding machine and then injected into the mold, where it cools and solidifies into the final part.

Injection molding is used to produce thin-walled plastic parts for a wide variety of applications, one of the most common being plastic housings. Plastic housing is a thin-walled enclosure, often requiring many ribs and bosses on the interior. These housings are used in a variety of products including household appliances, consumer electronics, power tools, and as automotive dashboards. Other common thin-walled products include different types of open containers, such as buckets. Injection molding is also used to produce several everyday items such as toothbrushes or small plastic toys. Many medical devices, including valves and syringes, are manufactured using injection molding as well.

## Capabilities

| | Typical | Feasible |
|---|---|---|
| Shapes: | Thin-walled: Cylindrical<br>Thin-walled: Cubic<br>Thin-walled: Complex | Flat |
| Part size: | Envelope: 0.01 $in^3$ - 80 $ft^3$<br>Weight: 0.5 oz - 55 lb | |
| Materials: | Thermoplastics | Composites<br>Elastomer<br>Thermosets |
| Surface finish - Ra: | 4 - 16 μin | 1 - 32 μin |
| Tolerance: | ± 0.008 in. | ± 0.002 in. |
| Max wall thickness: | 0.03 - 0.25 in. | 0.015 - 0.5 in. |
| Quantity: | 10000 - 1000000 | 1000 - 1000000 |
| Lead time: | Months | Weeks |
| Advantages: | Can form complex shapes and fine details<br>Excellent surrface finish<br>Good dimensional accuracy<br>High production rate<br>Low labor cost<br>Scrap can be recycled | |
| Disadvantages: | Limited to thin walled parts<br>High tooling and equipment cost<br>Long lead time possible | |
| Applications: | Housings, containers, caps, fittings | |

## Process Cycle

The process cycle for injection molding is very short, typically between 2 seconds and 2 minutes, and consists of the following four stages:

1. Clamping - Prior to the injection of the material into the mold, the two halves of the mold must first be securely closed by the clamping unit. Each half of the mold is attached to the injection molding machine and one half is allowed to slide. The hydraulically powered clamping unit pushes the mold halves together and exerts sufficient force to keep the mold securely closed while the material is injected. The time required to close and clamp the mold is dependent upon the machine - larger machines (those with greater clamping forces) will require more time. This time can be estimated from the dry cycle time of the machine.

2. Injection- The raw plastic material, usually in the form of pellets, is fed into the injection molding machine, and advanced towards the mold by the injection unit. During this process, the material is melted by heat and pressure. The molten plastic is then injected into the mold very quickly and the buildup of pressure packs and holds the material. The amount of material that is injected is referred to as the shot. The injection time is difficult to calculate accurately due to the complex and changing flow of the molten plastic into the mold. However, the injection time can be estimated by the shot volume, injection pressure, and injection power.

3. Cooling- The molten plastic that is inside the mold begins to cool as soon as it makes contact with the interior mold surfaces. As the plastic cools, it will solidify into the shape of the desired part. However, during cooling some shrinkage of the part may occur. The packing of material in the injection stage allows additional material to flow into the mold and reduce the amount of visible shrinkage. The mold can not be opened until the required cooling time has elapsed. The cooling time can be estimated from several thermodynamic properties of the plastic and the maximum wall thickness of the part.

4. Ejection- After sufficient time has passed, the cooled part may be ejected from the mold by the ejection system, which is attached to the rear half of the mold. When the mold is opened, a mechanism is used to push the part out of the mold. Force must be applied to eject the part because during cooling the part shrinks and adheres to the mold. In order to facilitate the ejection of the part, a mold release agent can be sprayed onto the surfaces of the mold cavity prior to injection of the material. The time that is required to open the mold and eject the part can be estimated from the dry cycle time of the machine and should include time for the part to fall free of the mold. Once the part is ejected, the mold can be clamped shut for the next shot to be injected.

After the injection molding cycle, some post processing is typically required. During cooling, the material in the channels of the mold will solidify attached to the part. This excess material, along with any flash that has occurred, must be trimmed from the part, typically by using cutters. For some types of material, such as thermoplastics, the scrap material that results from this trimming can be recycled by being placed into a plastic grinder, also called regrind machines or granulators, which regrinds the scrap material into pellets. Due to some degradation of the material properties, the regrind must be mixed with raw material in the proper regrind ratio to be reused in the injection molding process.

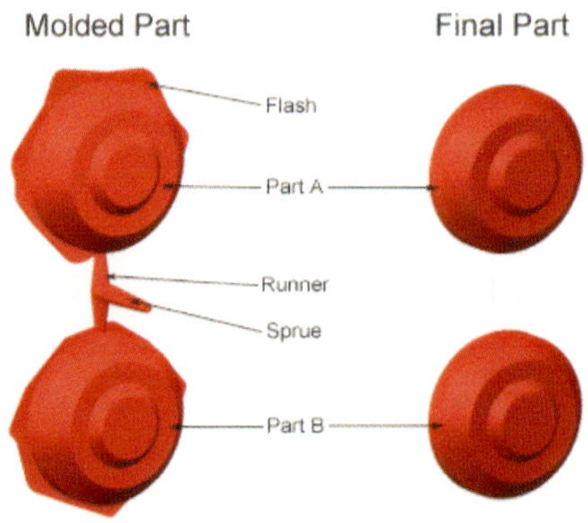


Injection molded part

## Equipment

Injection molding machines have many components and are available in different configurations, including a horizontal configuration and a vertical configuration. However, regardless of their design, all injection molding machines utilize a power source, injection unit, mold assembly, and clamping unit to perform the four stages of the process cycle.

## Injection Unit

The injection unit is responsible for both heating and injecting the material into the mold. The first part of this unit is the hopper, a large container into which the raw plastic is poured. The hopper has an open bottom, which allows the material to feed into the barrel. The barrel contains the mechanism for heating and injecting the material into the mold. This mechanism is usually a ram injector or a reciprocating screw. A ram injector forces the material forward through a heated section with a ram or plunger that is usually hydraulically powered. Today, the more common technique is the use of a reciprocating screw. A reciprocating screw moves the material forward by both rotating and sliding axially, being powered by either a hydraulic or electric motor. The material enters the grooves of the screw from the hopper and is advanced towards the mold as the screw rotates. While it is advanced, the material is melted by pressure, friction, and additional heaters that surround the reciprocating screw. The molten plastic is then injected very quickly into the mold through the nozzle at the end of the barrel by the buildup of pressure and the forward action of the screw. This increasing pressure allows the material to be packed and forcibly held in the mold. Once the material has solidified inside the mold, the screw can retract and fill with more material for the next shot.

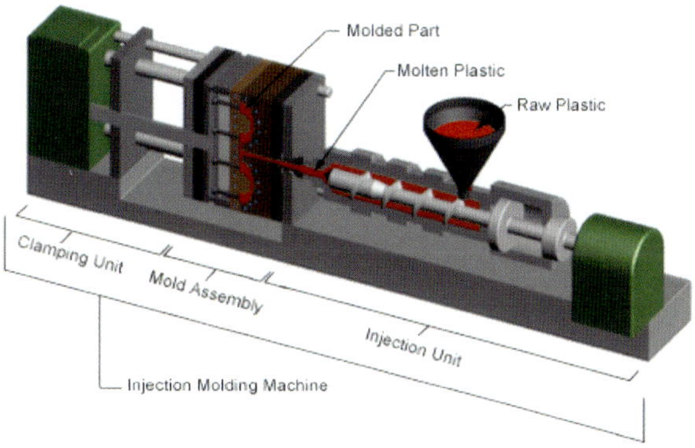


Injection molding machine - Injection unit

## Clamping Unit

Prior to the injection of the molten plastic into the mold, the two halves of the mold must first be securely closed by the clamping unit. When the mold is attached to the injection molding machine, each half is fixed to a large plate, called a platen. The front half of the mold, called the mold cavity, is mounted to a stationary platen and aligns with the nozzle of the injection unit. The rear half of the mold, called the mold core, is mounted to a movable platen, which slides along the tie bars. The hydraulically powered clamping motor actuates clamping bars that push the moveable platen towards the stationary platen and exert sufficient force to keep the mold securely closed while the material is injected and subsequently cools. After the required cooling time, the mold is then opened by the clamping motor. An ejection system, which is attached to the rear half of the mold, is actuated by the ejector bar and pushes the solidified part out of the open cavity.

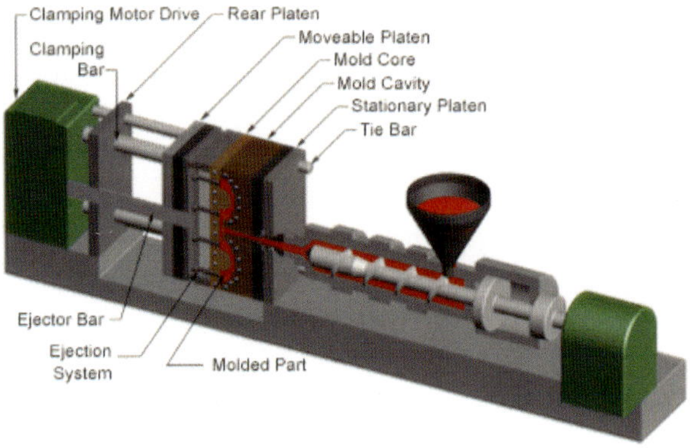


Injection molding machine - Clamping unit

## Machine Specifications

Injection molding machines are typically characterized by the tonnage of the clamp force they provide. The required clamp force is determined by the projected area of the parts in the mold and the pressure with which the material is injected. Therefore, a larger part will require a larger clamping force. Also, certain materials that require high injection pressures may require higher tonnage machines. The size of the part must also comply with other machine specifications, such as shot capacity, clamp stroke, minimum mold thickness, and platen size.

Injection molded parts can vary greatly in size and therefore require these measures to cover a very large range. As a result, injection molding machines are designed to each accommodate a small range of this larger spectrum of values. Sample specifications are shown below for three different models (Babyplast, Powerline, and Maxima) of injection molding machine that are manufactured by Cincinnati Milacron.

| | Babyplast | Powerline | Maxima |
|---|---|---|---|
| Clamp force (ton) | 6.6 | 330 | 4400 |
| Shot capacity (oz.) | 0.13 - 0.50 | 8 - 34 | 413 - 1054 |
| Clamp stroke (in.) | 4.33 | 23.6 | 133.8 |
| Min. mold thickness (in.) | 1.18 | 7.9 | 31.5 |
| Platen size (in.) | 2.95 x 2.95 | 40.55 x 40.55 | 122.0 x 106.3 |

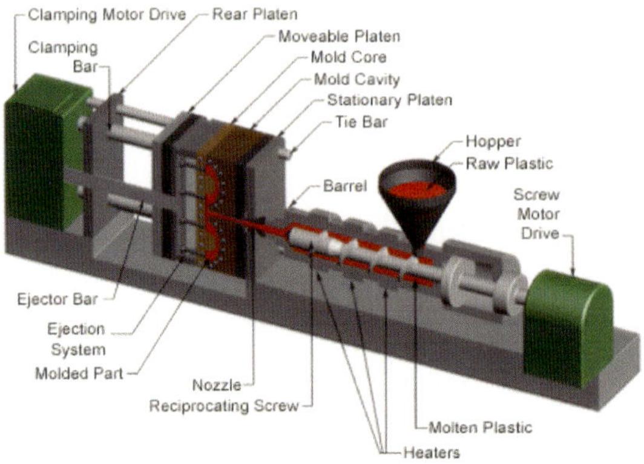


Injection molding machine

## Tooling

The injection molding process uses molds, typically made of steel or aluminum, as the custom tooling. The mold has many components, but can be split into two halves. Each half is attached inside the injection molding machine and the rear half is allowed to slide so that the mold can be opened and closed along the mold's parting line. The two main components of the mold are the mold core and the mold cavity. When the mold is closed, the space between the mold core and the mold cavity forms the part cavity, that will be filled with molten plastic to create the desired part. Multiple-cavity molds are sometimes used, in which the two mold halves form several identical part cavities.

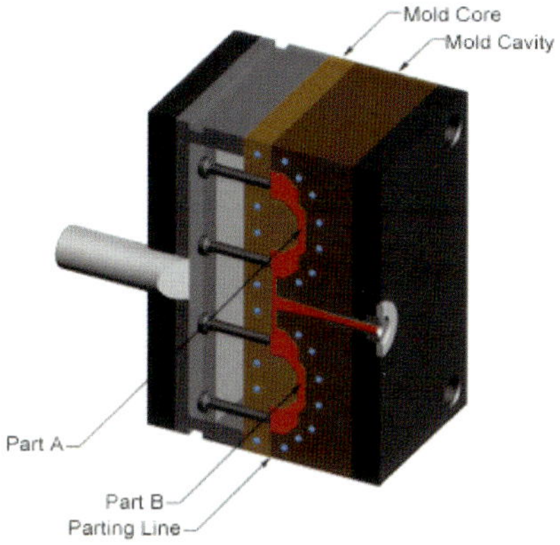


Mold overview

## Mold Base

The mold core and mold cavity are each mounted to the mold base, which is then fixed to the platens inside the injection molding machine. The front half of the mold base includes a support plate, to which the mold cavity is attached, the sprue bushing, into which the material will flow from the nozzle, and a locating ring, in order to align the mold base with the nozzle. The rear half of the mold base includes the ejection system, to which the mold core is attached, and a support plate. When the clamping unit separates the mold halves, the ejector bar actuates the ejection system. The ejector bar pushes the ejector plate forward inside the ejector box, which in turn pushes the ejector pins into the molded part. The ejector pins push the solidified part out of the open mold cavity.

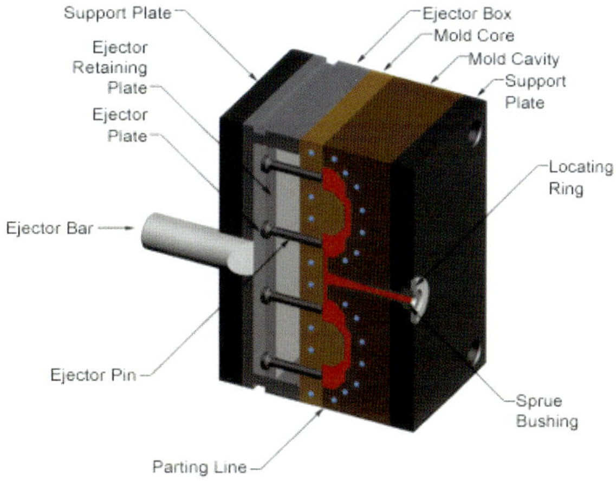


Mold Base

## Mold Channels

In order for the molten plastic to flow into the mold cavities, several channels are integrated into the mold design. First, the molten plastic enters the mold through the sprue. Additional channels, called runners, carry the molten plastic from the sprue to all of the cavities that must be filled. At the end of each runner, the molten plastic enters the cavity through a gate which directs the flow. The molten plastic that solidifies inside these runners is attached to the part and must be separated after the part has been ejected from the mold. However, sometimes hot runner systems are used which independently heat the channels, allowing the contained material to be melted and detached from the part. Another type of channel that is built into the mold is cooling channels. These channels allow water to flow through the mold walls, adjacent to the cavity, and cool the molten plastic.

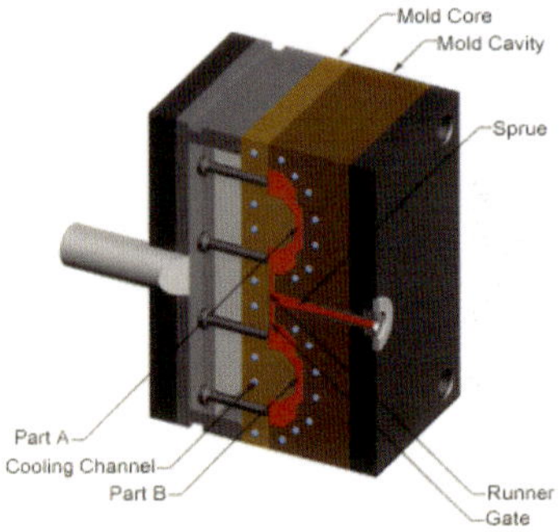


Mold Channels

## Mold Design

In addition to runners and gates, there are many other design issues that must be considered in the design of the molds. Firstly, the mold must allow the molten plastic to flow easily into all of the cavities. Equally important is the removal of the solidified part from the mold, so a draft angle must be applied to the mold walls. The design of the mold must also accommodate any complex features on the part, such as undercuts or threads, which will require additional mold pieces. Most of these devices slide into the part cavity through the side of the mold, and are therefore known as slides, or side-actions. The most common type of side-action is a side-core which enables an external undercut to be molded. Other devices enter through the end of the mold along the parting direction, such as internal core lifters, which can form an internal undercut. To mold threads into the part, an unscrewing device is needed, which can rotate out of the mold after the threads have been formed.

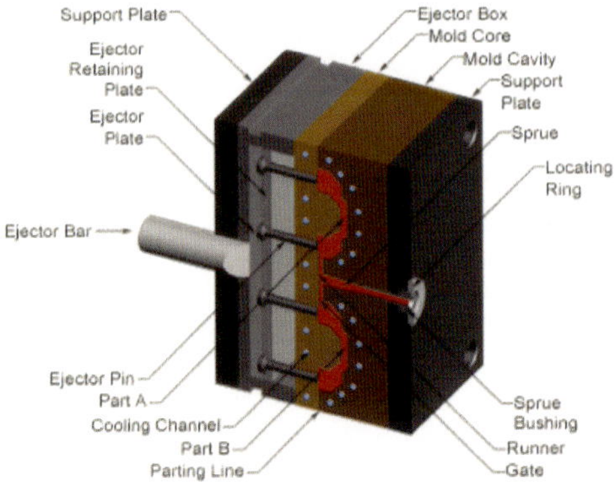

## Mold – Closed

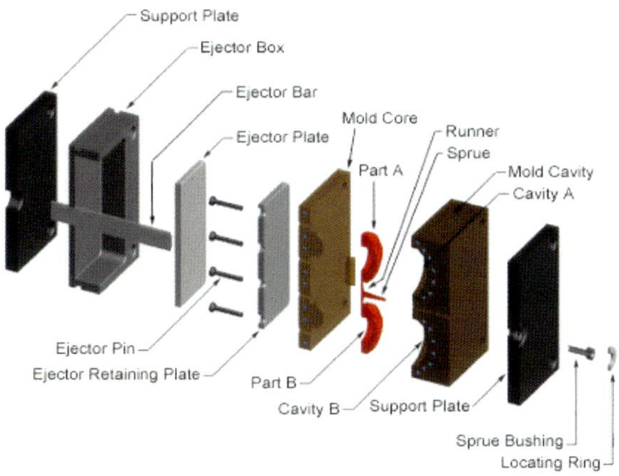


Mold - Exploded view

## Materials

There are many types of materials that may be used in the injection molding process. Most polymers may be used, including all thermoplastics, some thermosets, and some elastomers. When these materials are used in the injection molding process, their raw form is usually small pellets or a fine powder. Also, colorants may be added in the process to control the color of the final part. The selection of a material for creating injection molded parts is not solely based upon the desired characteristics of the final part. While each material has different properties that will affect the strength and function of the final part, these properties also dictate the parameters used in processing these materials. Each material requires a different set of processing parameters in the injection molding process, including the injection temperature, injection pressure, mold temperature, ejection temperature, and cycle time. A comparison of some commonly used materials is shown below.

| Material name | Abbreviation | Trade names | Description | Applications |
|---|---|---|---|---|
| Acetal | POM | Celcon, Delrin, Hostaform, Lucel | Strong, rigid, excellent fatigue resistance, excellent creep resistance, chemical resistance, moisture resistance, naturally opaque white, low/medium cost | Bearings, cams, gears, handles, plumbing components, rollers, rotors, slide guides, valves |
| Acrylic | PMMA | Diakon, Oroglas, Lucite, Plexiglas | Rigid, brittle, scratch resistant, transparent, optical clarity, low/medium cost | Display stands, knobs, lenses, light housings, panels, reflectors, signs, shelves, trays |
| Acrylonitrile Butadiene Styrene | ABS | Cycolac, Magnum, Novodur, Terluran | Strong, flexible, low mold shrinkage (tight tolerances), chemical resistance, electroplating capability, naturally opaque, low/medium cost | Automotive (consoles, panels, trim, vents), boxes, gauges, housings, inhalors, toys |

| | | | | |
|---|---|---|---|---|
| Cellulose Acetate | CA | Dexel, Cellidor, Setilithe | Tough, transparent, high cost | Handles, eyeglass frames |
| Polyamide 6 (Nylon) | PA6 | Akulon, Ultramid, Grilon | High strength, fatigue resistance, chemical resistance, low creep, low friction, almost opaque/white, medium/high cost | Bearings, bushings, gears, rollers, wheels |
| Polyamide 6/6 (Nylon) | PA6/6 | Kopa, Zytel, Radilon | High strength, fatigue resistance, chemical resistance, low creep, low friction, almost opaque/white, medium/high cost | Handles, levers, small housings, zip ties |
| Polyamide 11+12 (Nylon) | PA11+12 | Rilsan, Grilamid | High strength, fatigue resistance, chemical resistance, low creep, low friction, almost opaque to clear, very high cost | Air filters, eyeglass frames, safety masks |
| Polycarbonate | PC | Calibre, Lexan, Makrolon | Very tough, temperature resistance, dimensional stability, transparent, high cost | Automotive (panels, lenses, consoles), bottles, containers, housings, light covers, reflectors, safety helmets and shields |
| Polyester - Thermoplastic | PBT, PET | Celanex, Crastin, Lupox, Rynite, Valox | Rigid, heat resistance, chemical resistance, medium/high cost | Automotive (filters, handles, pumps), bearings, cams, electrical components (connectors, sensors), gears, housings, rollers, switches, valves |
| Polyether Sulphone | PES | Victrex, Udel | Tough, very high chemical resistance, clear, very high cost | Valves |
| Polyetheretherketone | PEEKEEK | | Strong, thermal stability, chemical resistance, abrasion resistance, low moisture absorption | Aircraft components, electrical connectors, pump impellers, seals |
| Polyetherimide | PEI | Ultem | Heat resistance, flame resistance, transparent (amber color) | Electrical components (connectors, boards, switches), covers, sheilds, surgical tools |
| Polyethylene - Low Density | LDPE | Alkathene, Escorene, Novex | Lightweight, tough and flexible, excellent chemical resistance, natural waxy appearance, low cost | Kitchenware, housings, covers, and containers |

| | | | | |
|---|---|---|---|---|
| Polyethylene - High Density | HDPE | Eraclene, Hostalen, Stamylan | Tough and stiff, excellent chemical resistance, natural waxy appearance, low cost | Chair seats, housings, covers, and containers |
| Polyphenylene Oxide | PPO | Noryl, Thermocomp, Vamporan | Tough, heat resistance, flame resistance, dimensional stability, low water absorption, electroplating capability, high cost | Automotive (housings, panels), electrical components, housings, plumbing components |
| Polyphenylene Sulphide | PPS | Ryton, Fortron | Very high strength, heat resistance, brown, very high cost | Bearings, covers, fuel system components, guides, switches, and shields |
| Polypropylene | PP | Novolen, Appryl, Escorene | Lightweight, heat resistance, high chemical resistance, scratch resistance, natural waxy appearance, tough and stiff, low cost. | Automotive (bumpers, covers, trim), bottles, caps, crates, handles, housings |
| Polystyrene - General purpose | GPPS | Lacqrene, Styron, Solarene | Brittle, transparent, low cost | Cosmetics packaging, pens |
| Polystyrene - High impact | HIPS | Polystyrol, Kostil, Polystar | Impact strength, rigidity, toughness, dimensional stability, naturally translucent, low cost | Electronic housings, food containers, toys |
| Polyvinyl Chloride - Plasticised | PVC | Welvic, Varlan | Tough, flexible, flame resistance, transparent or opaque, low cost | Electrical insulation, housewares, medical tubing, shoe soles, toys |
| Polyvinyl Chloride - Rigid | UPVC | Polycol, Trosiplast | Tough, flexible, flame resistance, transparent or opaque, low cost | Outdoor applications (drains, fittings, gutters) |
| Styrene Acrylonitrile | SAN | Luran, Arpylene, Starex | Stiff, brittle, chemical resistance, heat resistance, hydrolytically stable, transparent, low cost | Housewares, knobs, syringes |
| Thermoplastic Elastomer/Rubber | TPE/R | Hytrel, Santoprene, Sarlink | Tough, flexible, high cost | Bushings, electrical components, seals, washers |

## Possible Defects

| Defect | Causes |
|---|---|
| Flash | • Injection pressure too high<br>• Clamp force too low |
| Warping | • Non-uniform cooling rate |
| Bubbles | • Injection temperature too high<br>• Too much moisture in material<br>• Non-uniform cooling rate |
| Unfilled sections | • Insufficient shot volume<br>• Flow rate of material too low |
| Sink marks | • Injection pressure too low<br>• Non-uniform cooling rate |
| Ejector marks | • Cooling time too short<br>• Ejection force too high |

Many of the above defects are caused by a non-uniform cooling rate. A variation in the cooling rate can be caused by non-uniform wall thickness or non-uniform mold temperature.

## Design Rules

### Maximum Wall Thickness

- Decrease the maximum wall thickness of a part to shorten the cycle time (injection time and cooling time specifically) and reduce the part volume

| INCORRECT | CORRECT |
|---|---|
| 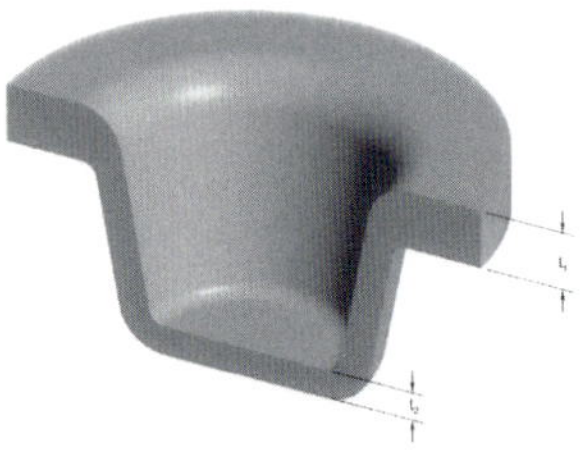<br>Part with thick walls | 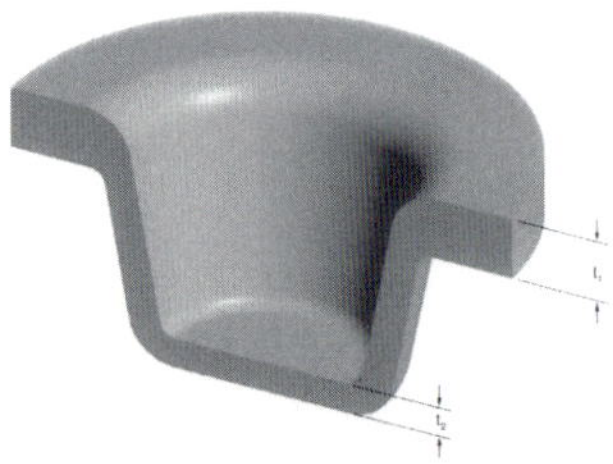<br>Part redesigned with thin walls |

- Uniform wall thickness will ensure uniform cooling and reduce defects

| INCORRECT | CORRECT |
|---|---|
| 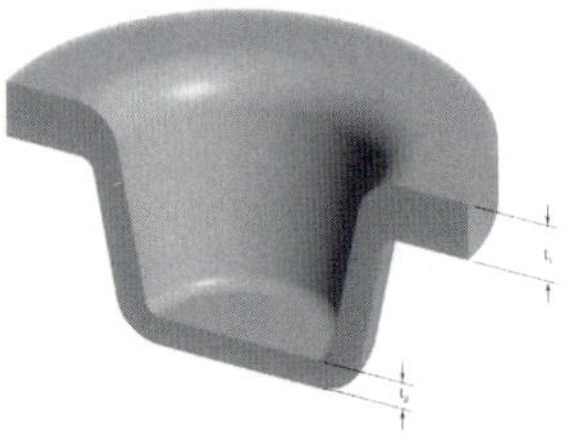<br>Non-uniform wall thickness ($t_1 \neq t_2$) | 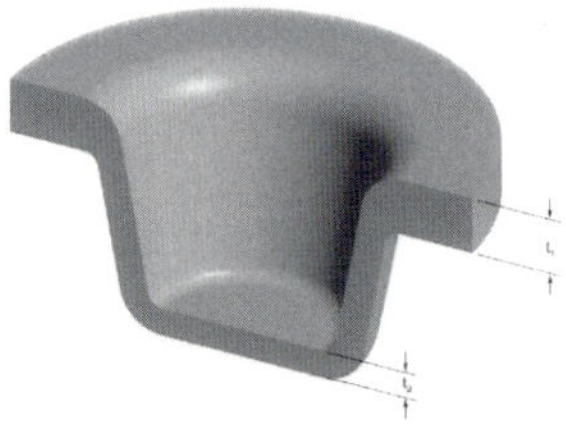<br>Uniform wall thickness ($t_1 = t_2$) |

## Corners

- Round corners to reduce stress concentrations and fracture
- Inner radius should be at least the thickness of the walls

| INCORRECT | CORRECT |
|---|---|
| 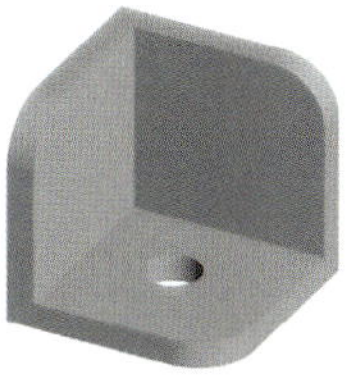 | 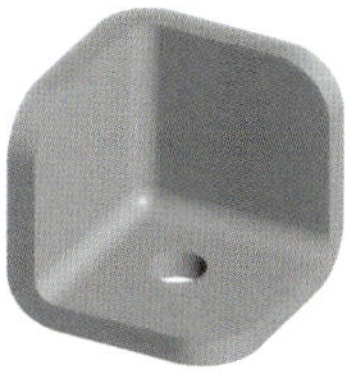 |
| Sharp corner | Rounded corner |

## Draft

- Apply a draft angle of 1° - 2° to all walls parallel to the parting direction to facilitate removing the part from the mold.

| INCORRECT | CORRECT |
|---|---|
|  | 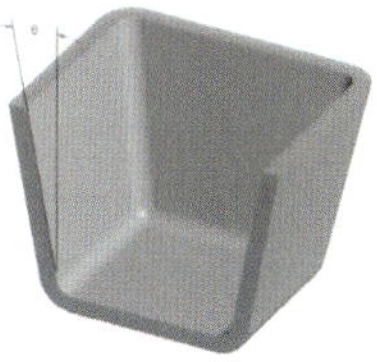 |
| No draft angle | Draft angle (θ) |

## Ribs

- Add ribs for structural support, rather than increasing the wall thickness

| INCORRECT | CORRECT |
|---|---|
| 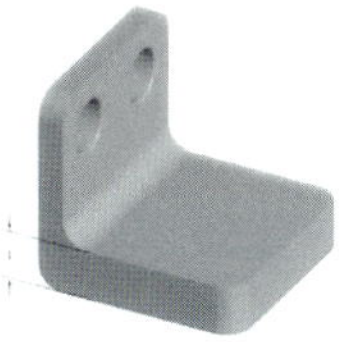 | 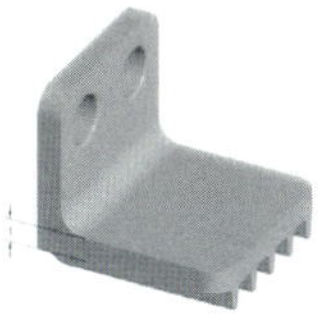 |
| Thick wall of thickness t | Thin wall of thickness t with ribs |

- Orient ribs perpendicular to the axis about which bending may occur

| INCORRECT | CORRECT |
|---|---|
| 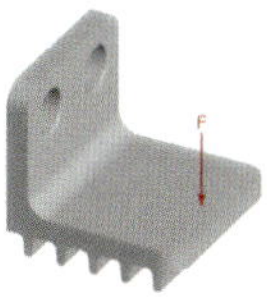 | 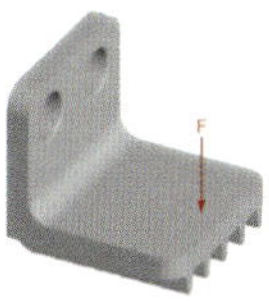 |
| Incorrect rib direction under load F | Correct rib direction under load F |

- Thickness of ribs should be 50-60% of the walls to which they are attached
- Height of ribs should be less than three times the wall thickness
- Round the corners at the point of attachment
- Apply a draft angle of at least 0.25°

INCORRECT

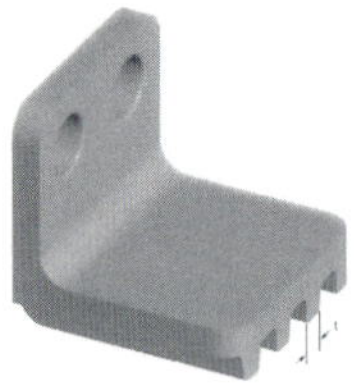

Thick rib of thickness t

CORRECT

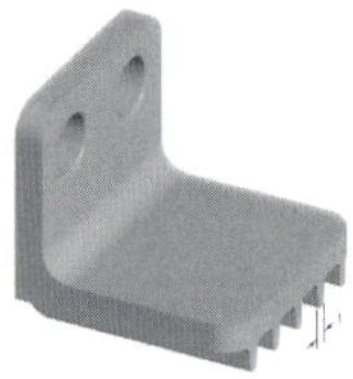

Thin rib of thickness t

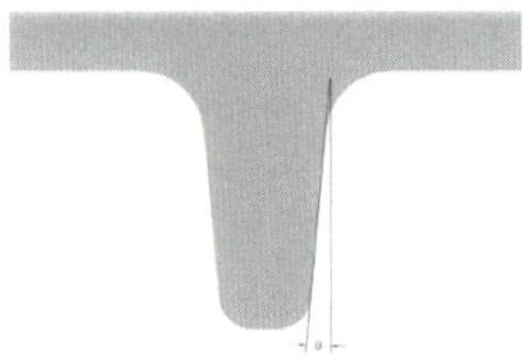

Close up of ribs

## Bosses

- Wall thickness of bosses should be no more than 60% of the main wall thickness
- Radius at the base should be at least 25% of the main wall thickness
- Should be supported by ribs that connect to adjacent walls or by gussets at the base.

INCORRECT

Isolated boss

CORRECT

Isolated boss with ribs (left) or gussets (right)

- If a boss must be placed near a corner, it should be isolated using ribs.

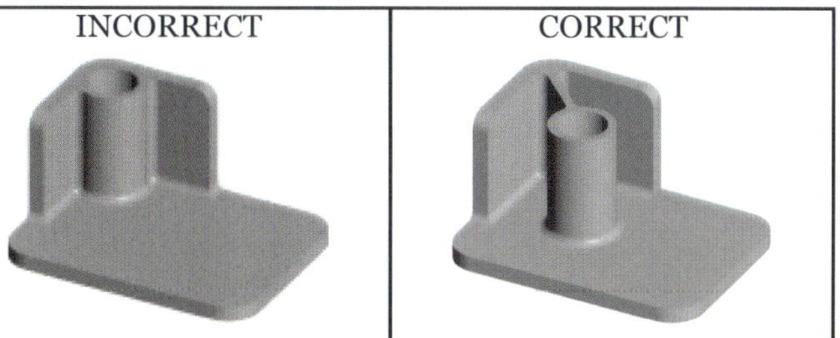


Boss in corner | Ribbed boss in corner

## Undercuts

- Minimize the number of external undercuts
- External undercuts require side-cores which add to the tooling cost
- Some simple external undercuts can be molded by relocating the parting line

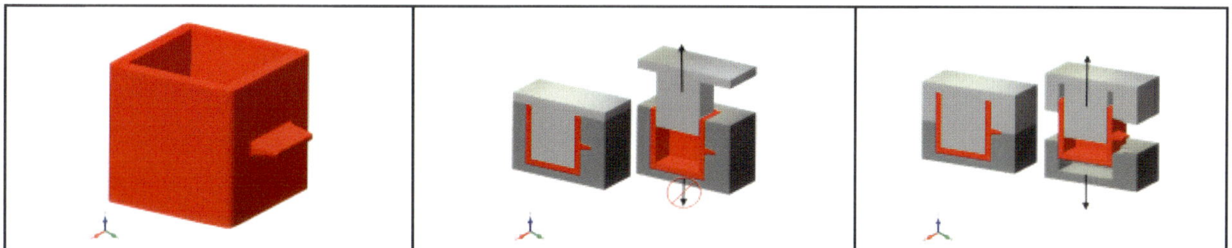

Simple external undercut | Mold cannot separate | New parting line allows undercut

- Redesigning a feature can remove an external undercut

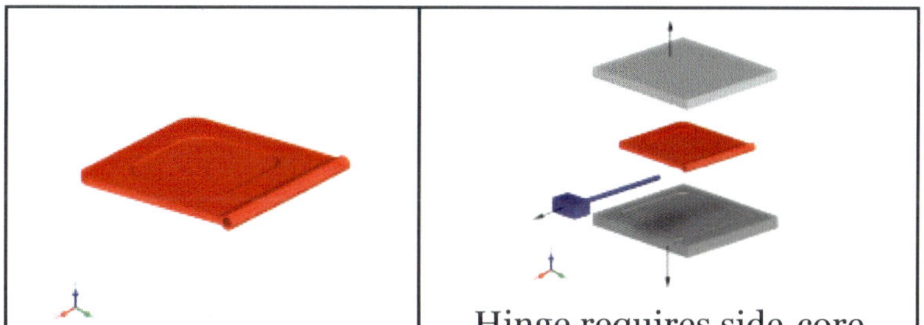

Part with hinge | Hinge requires side-core

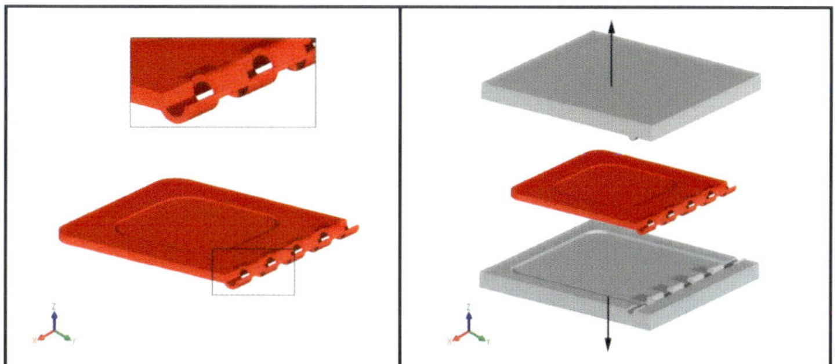

Redesigned hinge | New hinge can be molded

- Minimize the number of internal undercuts

- Internal undercuts often require internal core lifters which add to the tooling cost
- Designing an opening in the side of a part can allow a side-core to form an internal undercut

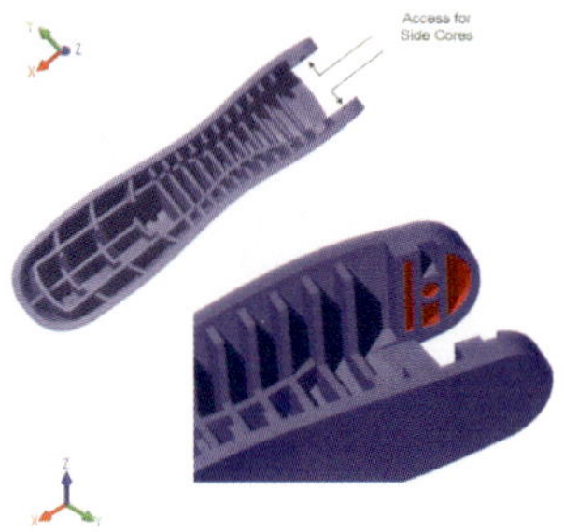

Internal undercut accessible from the side

- Redesigning a part can remove an internal undercut

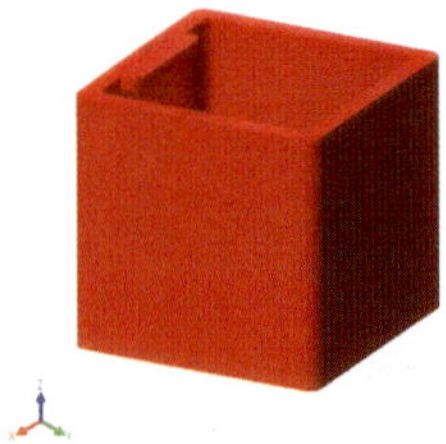

Part with internal undercut

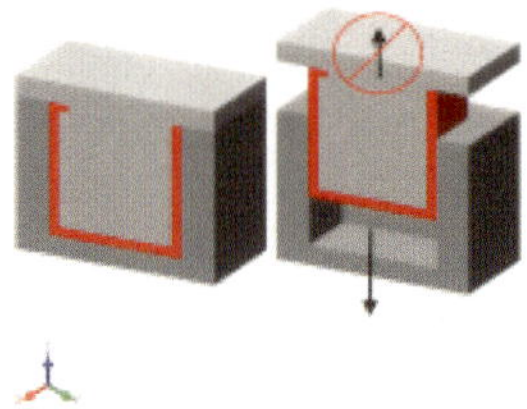

Mold cannot separate

Part redesigned with slot

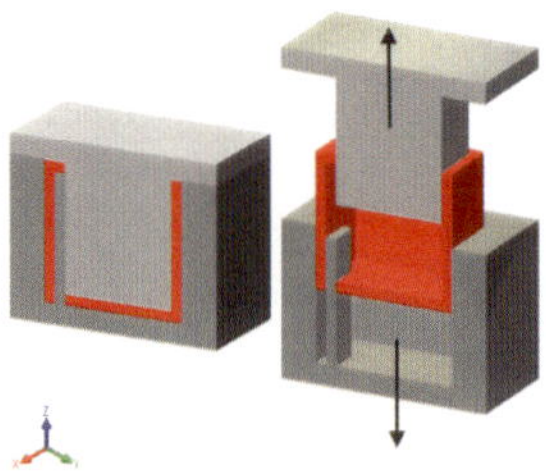

New part can be molded

- Minimize number of side-action directions
  - Additional side-action directions will limit the number of possible cavities in the mold

## Threads

- If possible, features with external threads should be oriented perpendicular to the parting direction.
- Threaded features that are parallel to the parting direction will require an unscrewing device, which greatly adds to the tooling cost.

## Cost Drivers

### Material Cost

The material cost is determined by the weight of material that is required and the unit price of that material. The weight of material is clearly a result of the part volume and material density; however, the part's maximum wall thickness can also play a role. The weight of material that is required includes the material that fills the channels of the mold. The size of those channels, and hence the amount of material, is largely determined by the thickness of the part.

### Production Cost

The production cost is primarily calculated from the hourly rate and the cycle time. The hourly rate is proportional to the size of the injection molding machine being used, so it is important to understand how the part design affects machine selection. Injection molding machines are typically referred to by the tonnage of the clamping force they provide. The required clamping force is determined by the projected area of the part and the pressure with which the material is injected. Therefore, a larger part will require a larger clamping force, and hence a more expensive machine. Also, certain materials that require high injection pressures may require higher tonnage machines. The size of the part must also comply with other machine specifications, such as clamp stroke, platen size, and shot capacity.

The cycle time can be broken down into the injection time, cooling time, and resetting time. By reducing any of these times, the production cost will be lowered. The injection time can be decreased by reducing the maximum wall thickness of the part and the part volume. The cooling time is also decreased for lower wall thicknesses, as they require less time to cool all the way through. Several thermodynamic properties of the material also affect the cooling time. Lastly, the resetting time depends on the machine size and the part size. A larger part will require larger motions from the machine to open, close, and eject the part, and a larger machine requires more time to perform these operations.

### Tooling Cost

The tooling cost has two main components - the mold base and the machining of the cavities. The cost of the mold base is primarily controlled by the size of the part's envelope. A larger part requires a larger, more expensive, mold base. The cost of machining the cavities is affected by nearly every aspect of the part's geometry. The primary cost driver is the size of the cavity that must be machined, measured by the projected area of the cavity (equal to the projected area of the part and projected holes) and its depth. Any other elements that will require additional machining time will add to the cost, including the feature count, parting surface, side-cores, lifters, unscrewing devices, tolerance, and surface roughness.

The quantity of parts also impacts the tooling cost. A larger production quantity will require a higher class mold that will not wear as quickly. The stronger mold material results in a higher mold base cost and more machining time.

One final consideration is the number of side-action directions, which can indirectly affect the cost. The additional cost for side-cores is determined by how many are used. However, the number of directions can restrict the number of cavities that can be included in the mold. For example, the mold for a part which requires 3 side-action directions can only contain 2 cavities. There is no direct cost added, but it is possible that the use of more cavities could provide further savings.

## Reaction Injection Molding

Reaction Injection Molding produces parts with intricate detail that are dimensionally stable, chemically resistant, physically tough, and wear resistant. This process is an excellent choice for larger plastic parts produced in short run or low volume production quantities.

- Polyurethane molded parts in 2-5 weeks
- Excellent flowability allows molding complex geometry including: louvers, ribs, bosses and openings
- Superior cosmetic appearance for visibly critical applications
- Superb encapsulation ability

The process takes its name from a chemical reaction that occurs within the tool. The plastics used are thermosets, either polyurethanes or foamed polyurethanes. The two components that produce the polyurethane are mixed just prior to injection into the tool. With the low viscosity and low injection pressures, large, complex parts can be produced economically in low quantities.

Considerable design freedom is possible, including thick and thin wall sections that are not good for injection molding, due to the uniform shrink characteristics. Foamed polyurethanes are natural thermal and acoustic insulators. Excellent flowability allows for the encapsulation of a variety of inserts.

Reaction injection molding is used in many industries for many types of parts. While bumpers for vehicles are produced in this process, most applications are for large, complex parts produced in quantities less than 5,000 units. Examples include panels for electrical equipment, enclosures for medical devices, and housings for computer and telecommunications equipment.

## Reaction Injection Molding Process

Reaction injection molding utilizes low-viscosity liquid polymers. Unlike traditional injection molding, rim always uses lightweight, cost-efficient thermoset polymers. Along with highly engineered resins, such as Dicyclopentadiene, otherwise known as DCPD.

The process begins with two liquids: isocyanate and polyol (for polyurethanes), which are stored in large tanks. High-pressure pumps circulate the liquids between the tanks and a multi-stream mix-head in a speedy, continuous loop. A piston then retracts inside the mix-head, breaking the loop and allowing the two liquids to combine.

The liquids are mixed at high velocity (1200 psi) using an impinging mixer before being transferred into the mold for the curing process. As liquid polymers require less pressure and lower temperatures to complete the part than traditional ingredients such as metals, RIM provides a further money-saving benefit via the use of lightweight machined aluminum molds.

The mixed polymer cures inside the mold using a combination of low pressure and heating of the mold to around 180 degrees Fahrenheit. Cure times within the mold vary from one to several minutes depending on the size, complexity and thickness of the part.

## Types of Reaction Injection Molding

RIM can be used to produce a number of physical characterics depending on the polymers chosen. Some common examples of these characteristics are solid, elastomeric, rigid foam, and flexible foam. This gives RIM a greater versatility in the type of parts it produced when compared not only to traditional injection molding but also to processes such as vacuum forming and cast molding.

Products created by RIM can be further enhanced via the addition of glass fibers to the liquid polymer, which is known as reinforced reaction injection molding (RRIM). Alternately, parts can be strengthened with a fiber mesh in a process called structural reaction injection molding (SRIM). These two variant processes can create parts that are even stronger and tougher, while still retaining the low weight and superior aesthetics of all RIM-created pieces.

## Advantages of Reaction Injection Molding

Polymer products created via reaction injection molding are typically known for being lightweight and flexible, especially when compared to more traditional materials such as steel, aluminum, or sheet molding compound (SMC). This lowered weight boasts a number of benefits.

A vehicle with bumpers or fenders made from polyurethane will be more fuel-efficient than one using heavier materials, saving the driver or company thousands of dollars in gas money. Machinery made from these lighter components will be easier to lift, use and handle. Launching your personal watercraft or installing your new spa tub will be a breeze with RIM-created polyurethane hulls, sides and other parts.

However, lighter weight isn't the only advantage of choosing rim. Thermoset polymer products are stable and flexible as well as strong, tough and durable, capable of resisting impact and wear without deforming. Wood rots, steel rusts, but polyurethane remains bright and shiny and retains its original shape after years of exposure to the elements.

At rim, all of our automobile and machinery parts are impact resisted and crash tested. In particular, help bumpers boast superior energy absorption when compared to other similar products.

In addition, reaction injection molded parts are non-corrosive and chemically resistant. This makes RIM the perfect choice for everything from heavy machinery to personal spas!

## Aesthetic Advantages

Unlike other processes, RIM allows for in-mold painting, producing eye-catching Class A finishes in either high or low gloss – right out of the mold. In addition to saving time and money usually needed for expensive, time-consuming post mold finishes, Romeo RIM's in-mold paintjobs are perfectly adhesive, and can last years without chipping or flaking away.

In-mold painting is not the only aesthetic benefit that comes with choosing RIM. RIM allows for the creation of molds with varied wall thickness – something that is completely impossible with traditional injection molding due to part shrinkage. RIM can be used to create far more intricate, detailed, complex and customizable parts than any other process available.

If that wasn't enough, RIM also provides the opportunity for adding B-side geometry, including louvers, ribs, bosses and opening. RIM molds allow for the encapsulation of a variety of inserts, further setting it apart from traditional injection molding, cast molding and vacuum forming.

RIM-produced products also make excellent thermal and acoustic insulators, further increasing their versatility and usefulness in a wide variety of industries.

## Compression Molding

One of several molding forms; compression molding is the act of using compression and heat to shape a raw material by means of a mold. In short, a raw material is heated until pliable, while the mold is closed for a certain time period. Upon removing the mold, the object may contain flash, excess product not conformed to the mold, which can be cut away.

The following factors must be considered when using a compression molding method:

- Material
- Shape
- Pressure
- Temperature
- Part thickness
- Cycle time

Plastics composed of both synthetic and natural materials are used in compression molding. Two types of raw plastics materials are most often used for compression molding:

- Thermoset plastics
- Thermoplastics

Thermoset plastics and thermoplastics are unique to the compression method of molding. Thermoset plastics refer to pliable plastics that once heated and set to a shape may not be changed, while thermoplastics harden as a result of being heated to a liquid state and then cooled. Thermoplastics can be reheated and cooled as much as necessary.

The amount of heat required and necessary instruments to produce the desired product vary. Some plastics require temperature in excess of 700 degrees F, while others in the low 200-degree range.

Time is also a factor. Material type, pressure, and part thickness are all factors which will determine how much time the part will need to be in the mold. For thermoplastics, the part and the mold will need to be cooled to an extent, so that piece being manufactured is rigid.

The force with which the object is compressed will depend on what the object can withstand, particularly in its heated state. For fiber reinforced composite parts being compression molded, the higher the pressure (force), often the better the consolidation of the laminate, and ultimately the stronger the part.

The mold used depends on the material and other objects used in the mold. The three most common types of molds used in compression molding of plastics are:

- Flash - requires accurate product inserted in the mold, removal of flash
- Straight-does not require accurate product, removal of flash
- Landed-requires accurate product, does not require removal of flash

It is important to ensure that no matter what material is used, the material covers all areas and crevices in the mold to ensure the most even distribution.

The process of compression molding begins with the material being placed into the mold. The product is heated until somewhat soft and pliable. A hydraulic tool presses the material against the mold. Once the material is set-hardened and has taken shape of the mold, an "ejector" releases the new shape. While some final products will require additional work, such as cutting away the flash, others will be ready immediately upon leaving the mold.

## Common Uses

Car parts and household appliances as well as clothing fasteners such as buckles and buttons are created with the help of compression molds. In FRP composites, body and vehicle armor is manufactured by means of compression molding.

Compression moulding is a traditional technique that uses a combination of heat and pressure to mould thermoset rubber, or plastic resin or into the customer's desired shape. The rubber or resin material is shaped using a heated mould tool and the clamping pressure of a press to force the flow of that material around inside the tool. The thermoset characteristic of the material is instigated by this process, and permanent chemical change takes place, fixing the shape of the product. Once this cure, or vulcanisation, process is complete, the mould is opened and the part extracted and allowed to cool, maintaining the shape of the mould that produced it, although the part may shrink during cooling to arrive at its final size.. The technique can be used to make a wide range of products such as wellington boots, machinery parts, seals, gaskets or even door stops and chair feet. This manufacturing process is very capable of reproducing the same product a large number of times, but it is not necessarily suitable for everything. Here are some of the advantages and disadvantages that relate to compression moulding as a process, to help you decide if it's right for you and your product.

Compression rubber mould tool

## Advantages of Compression Moulding

- Lower cost Tooling

Because the process does not involve an Injection or Transfer cycle, the tooling has less infrastructure requirements than tools designed for other moulding methods; other than the particular features of the cavity itself that will produce the moulded parts, which clearly have to replicate the detail of the product, there is little else needed in the way of additional features. Tools can be made of aluminium or lower cost grades of steel which can save cost, although any tool must be capable of withstanding the considerable moulding pressures required.

- Good for small production runs

The lower capital cost of manufacturing a mould tool, setting up a press and beginning to run production parts means that compression moulding is the most cost effective method of making smaller runs of parts. There is of course a breakeven point where the higher capital cost of an injection mould tool becomes viable due to the lower cost of the parts it makes, and making the correct determination of this point is a key consideration when developing the best production solution for a new product. No gates, sprues or runners This form of moulding does not use gates, sprues or runners which are tooling features that materials have to pass through in other methods of production before entering the mould cavity. These consume extra material and therefore cost, and can also detract from the cosmetic requirements of a part, both of which can be considerations when deciding the most appropriate tooling method for production.

- Good for large parts

This form of moulding is particularly suited to the production of large parts, which require significant bulk of material to manufacture. Because the material is loaded into the mould cavity directly, there is no limitation on the weight of part that can be made other than the size of press and tonnage required, whereas injection moulding presses are limited in the weight of part they can produce by the volume of the injection barrel that fills the mould.

## Disadvantages of Compression Moulding

Greater waste: Compression moulding is not such a precise method of making a product as injection moulding because the mould cavity has to be overfilled to some degree to achieve the correct pressure to cure the part. It is also often necessary to push air out of the cavity using excess material in order to create a void-free part, which again increases wastage whereas an injection mould is more precisely filled and inherently drives out the air as it is filled. Waste thermoset rubber or plastic cannot be melted down and reused so the cost of the part must reflect this. This production technique is essentially simple, but it does require more man power than a semi automatic injection moulding process to run. The skill level is often proportionately higher, and for larger parts or tools, manual handling can become an issue. Because the thermal conductivity of rubber and resin is relatively poor, bringing a blank of material up to curing temperature by contact with the hot surfaces of the mould does take time for larger parts, and this limits the speed at which parts can be cured and de-moulded. Injection moulding, however, shoots material into the mould that is already at a temperature just below curing, and so the extra time required to begin cure once the press closes is much shorter. Depending on the size of a part, a typical cycle may take around ten minutes, whereas the equivalent cycle of injection moulding would take about five minutes. This can make a significant difference when producing parts in higher volume, when production rate and part cost become the priority. While complex parts can be made by compression moulding, depending on the material required and the specifics of the design, it is generally better suited to the production of larger, simpler objects. This is because the limited flow of material within the cavity often makes the elimination of voids, air traps and knit lines difficult when trying to produce complex parts. It is much easier to produce clean, consistently coloured rubber parts by injection or transfer moulding, as the uncured blank of compound used in compression moulding can pick up specks of dirt which then simply get pressed and cured into the surface of the part, and result in poor cosmetic appearance despite the part being dimensionally and physically acceptable. This increases reject rates and waste. Flash is a necessary by-product of compression moulding due to the need to expel air during the moulding process by overfilling the mould cavity. This flash can also vary in thickness depending on the exact fill of each particular shot, leading to issues with variable dimensions of the part over the split line. The flash must be mechanically cut off and again is a thermoset material that cannot be recycled easily, and the resulting split line witness mark is often more evident than on an injection moulded part which may be an aesthetic problem to the customer. Generally, compression moulded parts require more labour effort to trim and finish them, which also increases the cost. The repetitive nature of this process means that the moulds themselves can become victims of general wear and tear, with manually loaded blanks of material and, often, the mould is made to be run loose and not bolted into a press. This can lead to deterioration of trim grooves and fine details, or in extreme cases, impact damage on mould faces caused by poor handling. Overall, the process of compression moulding is ideal when used appropriately and can help to keep overall project costs to a minimum, although individual part costs will be higher compared to other techniques. It is vital to consider all the required aspects of a product, its performance and its life cycle when designing the most appropriate production solution, as each process technique has a different combination of costs and benefits. The proper solution may not require an expensive injection mould tool, and thus even if the part cost is higher, overall, compression moulded solutions remain as relevant as ever.

# Rotational Molding

Rotational moulding is a manufacturing process used for processing inter alia Polyethylene or Polypropylene in plastic products. Examples are products like buoys, water and / or fuel tanks, toys, traffic guidance, fenders etc. The products of Pentas vary in their weight from 0.5 kg to 300 kg of plastic per product. The numbers of products produced per mould vary between 100 to 5,000 products.

## Working of Rotational Moulding Process

The rotational moulding process consists out of four phases. It all starts with filling the mould with powdered resin. The inside of this mould is the negative of the final product. Because each product is different, the moulds are specifically developed for each customer.

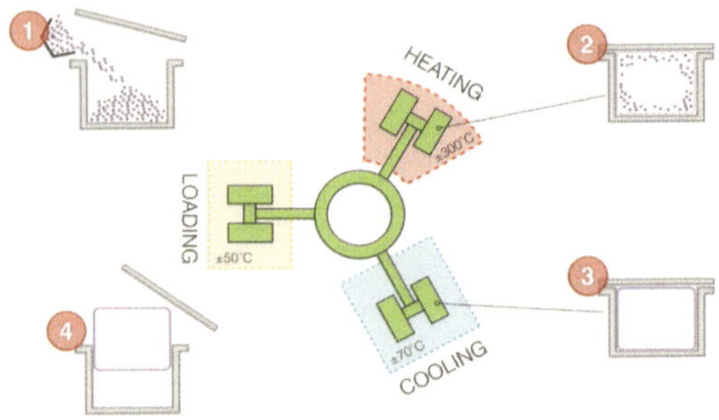


When the mould is closed it will be heated in the next phase. In this heating process, the mould will rotate around the horizontal as well as the vertical axis. During this heating process the plastic melts and attaches itself to the walls of the mould. When the powder has molten and it is formed around the walls of the mould subsequent the cooling process can be started. The cooling happens in the next phase in which large fans blows air through the mould walls. When the product is sufficiently solidified, it will be removed from the mould and the rotational moulding process is completed, there is now a produced plastic product.

## Typical Rotational Molding Applications

Rotational molding permits production of a countless number of fully or partially closed items. Design versatility of rotationally molded pieces is almost unlimited.

The rigidity or flexibility of an item is controlled by the properties of the resin used and by the wall thickness of the molding.

Some typical applications for which rotational molding is particularly suited include the following:

- Commercial, industrial and agricultural storage tanks ranging in size from 5 gallons to 22,000 gallons;
- Containers for packaging and material handling;
- A variety of industrial parts, especially covers and housings, water softening tanks, tote bins;

- Numerous under-the-hood and in-the-cab automotive parts;
- Hobby horses, pool tables and many other small or large toys with complex shapes;
- Baby cribs, balls, dolls, and doll parts;
- Display window mannequins and other hollow display figures of all sizes, and;
- Playground equipment and sporting equipment such as golf carts, footballs, juggling pins, helmets, dumbbells, golf tee markers, canoes and kayaks.

Rotomolded parts are also used in portable outhouses, battery cases, light globes, vacuum cleaner and scrubber housings and garbage containers. Furniture, game housings, surf boards, traffic barricades, display cases and ducting can also be produced by rotomolding. The list above indicates just some of the possibilities. Figures illustrate some of these applications.

## Design – Endless Potential

With rotomolding, a plastics product designer can create a huge array of innovative products. Typical design concerns that are handled routinely by rotational molding include the following:

- Maintaining uniform wall thickness: Rotational molding can provide a more consistently uniform wall thickness for a part compared to other plastic processing methods;
- Producing double-wall construction: Rotational molding can provide uniform double-wall construction on parts;
- Molding thicker corners: Due to the process, rotomolded parts will have thicker outer corners which help strengthen the parts; and
- Molding inserts, reinforcing ribs, kiss-off ribbing and undercuts: These are easily included in a roto-molded part.

## Advantages of Rotational Molding

Rotational molding offers significant advantages when compared with other molding techniques or thermoforming:

- Costs for molds and tooling are relatively low;
- The rotomolding technique is easily adapted to short production runs, particularly when sets of multiple-cavity molds are used;
- Hollow, totally enclosed items as well as pieces with openings can be made;
- Rotational molding eliminates the need for secondary tooling;
- There is little or no waste due to resin scrap;
- Wall thickness and piece weight can be easily controlled;
- Rotational molding procedures assure uniform wall thickness;

- Deviations can be controlled to within a maximum tolerance of ±10 percent;
- Pieces with intricate contours and undercuts can be easily molded;
- Virtually any size piece can be rotationally molded;
- There is a minimum of crosssectional deformation and warpage;
- Rotational molding yields pieces with excellent surface detail and finish;
- Rotationally molded items are virtually stress free;
- Identical or similar items or different sections on one piece can be molded at the same time in different colors on a single spindle;
- Plastic or metal inserts can often be molded as integral parts of the item; and
- Double wall constructions are feasible

# Thermoforming

Thermoforming is a process of heating a thermoplastic sheet to its softening point. The sheet is stretched across a single-sided mold and then manipulated. Then, it cools into the desired shape. The most common methods to get the sheet to conform to its final shape are vacuum-forming, pressure-forming and mechanical forming. Thermoforming has innumerable applications and can be used by several different industries. Here is a brief overview of thermoforming, its benefits, and applications.

## Key Terminology used

The process of thermoforming uses some specific terminology with which you should be familiar. Knowing the difference between these terms can be the difference between having a successful project and one that is lackluster.

- Thermoforming Thermoforming is a general term that refers to the process of transforming a plastic sheet into a 3-dimensional shape by using heat, vacuum and pressure.
- Thermoplastics Thermoplastics refers to a type of plastic made from polymer resins that become homogenized when heated.
- Vacuum Forming The specific process that involves forming a part by heating and stretching the plastic across a mold using a vacuum. Normally the mold is open and the force involved with forming the sheet is limited to about 15 psi.
- Pressure Forming This process adds a pressure box to the tooling package. It utilizes both vacuum and positive air pressure. This process generates as much as three to four times the forming pressure as vacuum forming does. Therefore, fine details such as surface textures can be formed on the mold without incurring excessive extra costs.

- Twin Sheet Forming As the term suggests, there are two molds for this process: one on the top and one on the bottom. Therefore, two sheets of plastic must be heated and formed at the same time. A fused joint must be placed around the perimeter of the mold while air pressure is injected between the sheets. This process is ideal for forming hollow parts that require a distinct upper and lower shape.

## Stages of Thermoforming

As like any process, there are several stages of thermoforming. The typical thermoforming process goes through the following stages:

- Design
- Prototyping
- Final tooling
- Production
- Shipping

## Equipment used for Thermoforming

The thermoforming process requires a few pieces of equipment. The forming process needs either pressure formers or vacuum formers. It also requires a three or four-station Rotary Transformer. The secondary operation of the process needs CNC robotic routers. Normally the bed for this piece of equipment does not exceed 60"X 120". The tooling process requires machine aluminum molds, cast aluminum molds, composite or temporary molds, or male/female molds.

## Advantages of Thermoforming

Thermoforming holds many advantages that make it a favorable process over other types of molding. Some of these advantages include:

- The ability to create several finished parts from the same material
- It allows for the detection of possible design and fit issues before it is too late
- It saves time and money
- It is beneficial when using large parts

## Benefits of Thermoforming

The thermoforming process offers many benefits to both the manufacturer and end user. Some of these benefits include:

- Efficient
- Cost-effective

- Short lead time
- Easy to add details
- Freedom of design

## Common Materials

One of the best aspects of thermoforming is its versatility. This feature can be seen in the vast number of materials that can be used for the process. Each material has applications and functions for which it works best.

- ABS (Acrylonitrile Butadiene Styrene) – This is a common material that has good stiffness and impact strength. It comes in almost any color as well as several textures.
- Acrylic (Polymethyl Methacrylate, Plexiglass or PMMA) – This material is clear and abrasion-resistant. It can be fabricated easily, is available in impact-modified grades and also comes in many colors.
- HDPE (High Density Polyethylene) – This material is resistant to impact as well as chemicals. It also has great cold-temperature properties. However, it is not as stable as many other available materials.
- HIPS (High-Impact Polystyrene) – This material has a low cost attached to it. IT also forms easily and is available in many colors. It is more brittle than some other materials, including ABS.
- HMPWE (High Molecular Weight Polyethylene) – This material has high impact strength, is chemical resistant and puncture resistant. It is ideal for medical joint replacement, recreational skis, and bottles.
- KYDEX (PMMA/PVC blend) – This is a good general-use all-purpose material. It is resistant to chemicals and heavy impacts. It is available in many colors and textures.
- LEXAN – This material has great variability, is flame-resistant, scratch-resistant, and can stand up to various types of weather.
- PC (Polycarbonate) – This material has incredibly high impact strength. It is clear and has a high temperature resistance.
- Pennite (glass-filled nylon) – This material is strong and stiff. It is also cost-effective, especially if it is used to replace metal.
- PEI (Polyetherimide Ultem) – This is very high-temperature grade material. It is autoclavable and has a natural amber color.
- PET (Polyethylene Terephthalate) – This material is commonly used food packaging. It is clear and has a low cost. It can be FDA certified if needed.
- PETG (Polyethylene Terephthalate Glycol) – This material is clear and has excellent impact strength. It also forms well. Medical-grade material is available upon need.

- PP (Polypropylene) – This material has excellent chemical resistance. It is rigid and has good impact strength. It is good at higher temperatures but dimensionally is not as stable as other materials.
- PVC (Polyvinyl Chloride) – This rigid material is strong and has good impact strength. It is also flame-retardant. However, its availability is limited.
- Royalite – This material is durable, has high impact strength and high tensile strength. It also has high and low temperature performance.
- RPET (Reprocessed Polyethylene Terephthalate) – This material is commonly used in food packaging. It is clear and has a low cost. It can be FDA certified if needed.
- TPO (Thermoplastic PolyOlefin) – This material has superior impact properties. It is available with a high-glass finish. It is good for outdoor application. However, it is difficult to form, particularly for applications that need deep draw shapes.
- Viny – This material is durable, flame-resistant, and a good conductor of electricity.
- Specialty materials – Other materials are available as well. You just need to ask your fabricator or guidance.

## Thermoform Applications

The thermoforming process can be used in a wide variety of applications as well. Some of the most common ones include but are not limited to:

- Retail clamshell packaging
- Packaging blisters
- Pick and place trays
- Material and handling trays and covers
- Shipping trays
- Medical packaging
- Pop displays
- Packaging inserts

## Tolerances

Typical tolerances for thermoforming processes include:

- Forming: Part to part +/- .025", depending on the material
- Perimeter Die Cut: +/- .060"

## Thin-gauge Tooling

Typical parameters for thin-gauge thermoforming tooling include:

- Maximum mold size – up to 30" wide
- Maximum depth of draw – 8.0"
- Maximum sheet thickness – .80"
- Molds are temperature controlled aluminum
- Trim is via steel rule die

## General Practices

For the thermoforming process to be completed accurately the following general practices should always be followed:

- Dimensions should be drawn to and from a controlled surface.
- The controlled surface should be against the mold.
- There may typically be a 2.0" callout for mold relief.
- There may be +/- 10% variance in thickness of stock material.
- Material stretch should be dependent on mold configuration and depth of draw.

## Thermoplastics

In essence, thermoplastics are the final products that result from the thermoforming process. A major benefit of thermoplastics is their tolerance to repeated activation. So, it can be reheated and reshaped several times. This characteristic also makes thermoplastics recyclable. Further, due to the chemistry involved, thermoplastic materials exhibit the same characteristics as rubber and can have the same strength as aluminum. The temperature tolerance of thermoplastic materials varies. Some can retain their properties up to 100 degrees F, while others can withstand temperatures up to 600 degrees F. Most thermoplastics have no known solvent at room temperature and also function well as both electrical and thermal insulation. They can also be electrically conductive if metal or carbon is added.

## Nature of Thermoplastics

Thermoplastics have a long history, dating back to the mid-1800s. They were originally used as a substitute for ivory and have grown to become a part of everyday life today. Some common places where you can see thermoplastics include:

- Sports equipment
- Toys
- Auto parts

- Compact disks
- Food storage containers
- • Eyeglass lenses
- Grocery bags
- Bulletproof vests

## Vacuum Forming

Thermoforming and vacuum forming are closely related. In fact, vacuum forming is a type of thermoforming. However, there are some important differences. So, it is important to be able to distinguish between the two. Vacuum forming requires some additional processes that cause plastic to conform to the mold during forming. Vacuum forming has more restrictive uses as well. It is normally reserved for applications that use shallow plastic parts that need to be formed into cavities. Other important information to know about vacuum forming includes:

## Suitable Materials

There are fewer materials that are suitable for vacuum forming applications. The best materials for this purpose include:

- Thermoplastics
- HIPS

## Applications

Vacuum forming is ideal for some specific applications because the vacuum-formed components can replace complex fabricated sheet metals, fiberglass, or plastic injection molding. Some common places where you can see evidence of vacuum forming include:

- Kiosks
- ATMs
- Medical imaging enclosures
- Diagnostic equipment

## Problems with Vacuum Forming

Despite all of its positive attributes, vacuum forming does has some drawbacks as well. Some of the most common problems include:

- The molds can easily absorb moisture forming bubbles within the inner layers
- Webs can form around the mold and negatively affect the final product
- Objects can stick to the mold

## References


- Types-molding-processes-7651143: sciencing.com, Retrieved 19 July 2018
- What-is-blow-molding: milacron.com, Retrieved 26 March 2018
- Transfer-molding: themetalcasting.com, Retrieved 12 May 2018
- What-is-reaction-injection-molding: romeorim.com, Retrieved 13 July 2018
- What-is-compression-molding-820345: thoughtco.com, Retrieved 30 April 2018
- Advantages-and-disadvantages-of-compression-moulding: martins-rubber.co.uk, Retrieved 27 March 2018
- What-is-thermoforming-and-how-can-the-process-be-used-in-product-manufacturing: radiofrequencywelding.com, Retrieved 11 May 2018

# 5

# Forming Processes

Forming processes are manufacturing processes which utilize different stresses like shear, tension, compression, etc. to cause plastic deformation in materials for producing desired shapes. This chapter aims to provide a comprehensive understanding of forming processes and includes vital topics such as rolling, extrusion, swaging, explosive forming, electroforming, hydroforming, etc.

Forming is a mechanical process used in manufacturing industries wherein materials (mostly metals) undergo plastic deformations and acquire required shapes and sizes by application of suitable stresses such as compression, shear and tension. In the forming process, no material is removed; it is completely displaced and deformed into the required shape. Some of the commonly used forming processes in the manufacturing industry are:

- Forging
- Rolling
- Extrusion
- Thread rolling
- Rotary swaging
- Explosive forming
- Electromagnetic forming

Forming, also known as "metal forming," includes a wide range of manufacturing processes in which metal is deformed into a required shape by the application of suitable stresses. To make the metal plastically deformed, forces must be applied that are greater than the yield strength of the metals. The magnitude of the compression, stretching or bending in the material, is directly proportional to the force applied.

In industrial processes, forming is characterized by:

- High levels of loads and stresses ranging from 50 to 2,500 newtons per square millimeter.
- Big, heavy and expensive machinery, which can generate a high amount of stresses and loads.
- Many parts that are produced in less time, which helps in maximizing the production economy or reaching economies of scale.

## Sheet Metal Forming

Sheet metal forming processes are those in which force is applied to a piece of sheet metal to modify its geometry rather than remove any material. The applied force stresses the metal beyond its yield strength, causing the material to plastically deform, but not to fail. By doing so, the sheet can be bent or stretched into a variety of complex shapes.

## Process

### Bending

Bending is a metal forming process in which a force is applied to a piece of sheet metal, causing it to bend at an angle and form the desired shape. A bending operation causes deformation along one axis, but a sequence of several different operations can be performed to create a complex part. Bent parts can be quite small, such as a bracket, or up to 20 feet in length, such as a large enclosure or chassis. A bend can be characterized by several different parameters, shown in the image below.

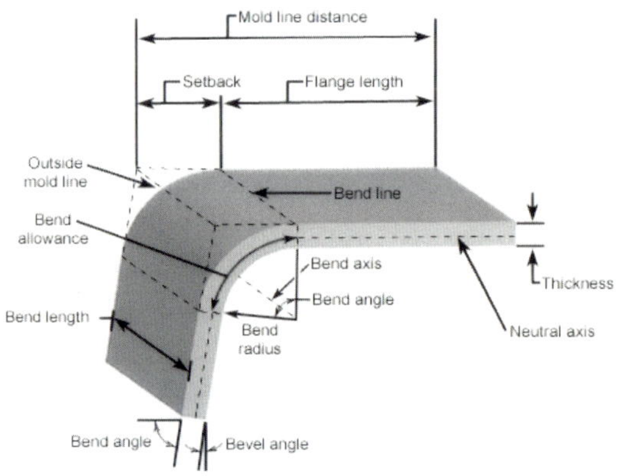


Bending Diagram

- Bend line - The straight line on the surface of the sheet, on either side of the bend, that defines the end of the level flange and the start of the bend.
- Outside mold line - The straight line where the outside surfaces of the two flanges would meet, were they to continue. This line defines the edge of a mold that would bound the bent sheet metal.
- Flange length - The length of either of the two flanges, extending from the edge of the sheet to the bend line.
- Mold line distance - The distance from either end of the sheet to the outside mold line.
- Setback - The distance from either bend line to the outside mold line. Also equal to the difference between the mold line distance and the flange length.
- Bend axis - The straight line that defines the center around which the sheet metal is bent.
- Bend length - The length of the bend, measured along the bend axis.
- Bend radius - The distance from the bend axis to the inside surface of the material, between the bend lines. Sometimes specified as the inside bend radius. The outside bend radius is equal to the inside bend radius plus the sheet thickness.

- Bend angle - The angle of the bend, measured between the bent flange and its original position, or as the included angle between perpendicular lines drawn from the bend lines.

- Bevel angle - The complimentary angle to the bend angle.

The act of bending results in both tension and compression in the sheet metal. The outside portion of the sheet will undergo tension and stretch to a greater length, while the inside portion experiences compression and shortens. The neutral axis is the boundary line inside the sheet metal, along which no tension or compression forces are present. As a result, the length of this axis remains constant. The changes in length to the outside and inside surfaces can be related to the original flat length by two parameters, the bend allowance and bend deduction, which are defined below.

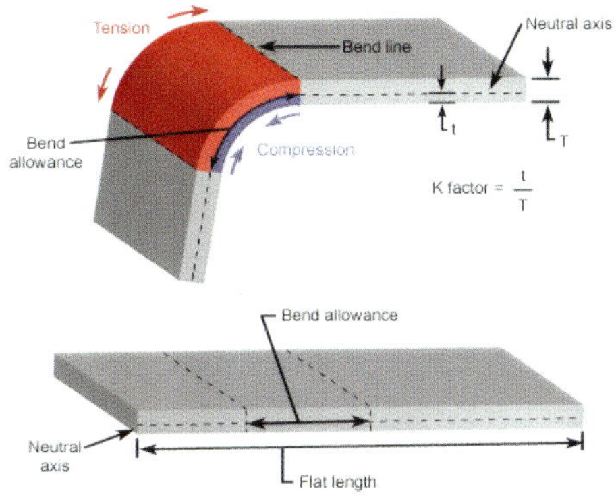


Neutral Axis

- Neutral axis - The location in the sheet that is neither stretched nor compressed, and therefore remains at a constant length.

- K-factor - The location of the neutral axis in the material, calculated as the ratio of the distance of the neutral axis (measured from the inside bend surface) to the material thickness. The K-factor is dependent upon several factors (material, bending operation, bend angle, etc.) and is typically greater than 0.25, but cannot exceed 0.50.

- Bend allowance - The length of the neutral axis between the bend lines, or in other words, the arc length of the bend. The bend allowance added to the flange lengths is equal to the total flat length.

- Bend deduction - Also called the bend compensation, the amount a piece of material has been stretched by bending. The value equals the difference between the mold line lengths and the total flat length.

When bending a piece of sheet metal, the residual stresses in the material will cause the sheet to springback slightly after the bending operation. Due to this elastic recovery, it is necessary to over-bend the sheet a precise amount to achieve the desired bend radius and bend angle. The final bend radius will be greater than initially formed and the final bend angle will be smaller. The ratio of the final bend angle to the initial bend angle is defined as the springback factor, KS. The amount of springback depends upon several factors, including the material, bending operation, and the initial bend angle and bend radius.

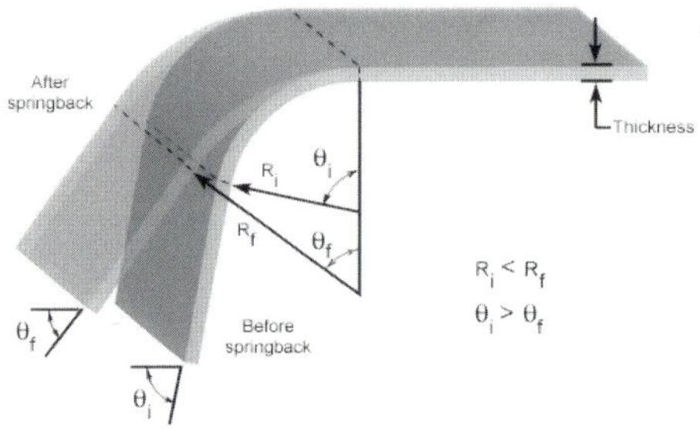


Springback

Bending is typically performed on a machine called a press brake, which can be manually or automatically operated. For this reason, the bending process is sometimes referred to as press brake forming. Press brakes are available in a range of sizes (commonly 20-200 tons) in order to best suit the given application. A press brake contains an upper tool called the punch and a lower tool called the die, between which the sheet metal is located. The sheet is carefully positioned over the die and held in place by the back gauge while the punch lowers and forces the sheet to bend. In an automatic machine, the punch is forced into the sheet under the power of a hydraulic ram. The bend angle achieved is determined by the depth to which the punch forces the sheet into the die. This depth is precisely controlled to achieve the desired bend. Standard tooling is often used for the punch and die, allowing a low initial cost and suitability for low volume production. Custom tooling can be used for specialized bending operations but will add to the cost. The tooling material is chosen based upon the production quantity, sheet metal material, and degree of bending. Naturally, a stronger tool is required to endure larger quantities, harder sheet metal, and severe bending operations. In order of increasing strength, some common tooling materials include hardwood, low carbon steel, tool steel, and carbide steel.

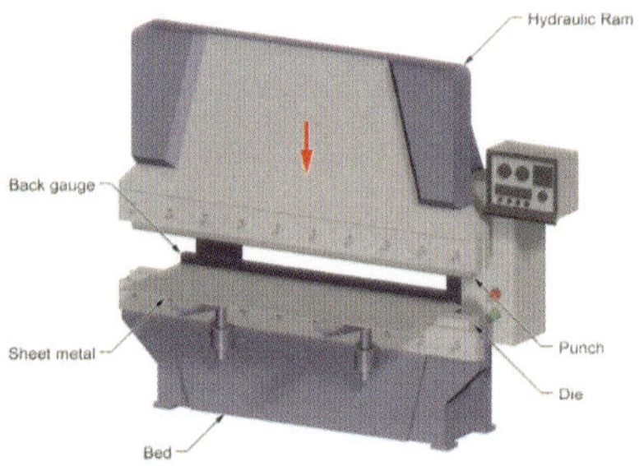


Press Brake (Open)

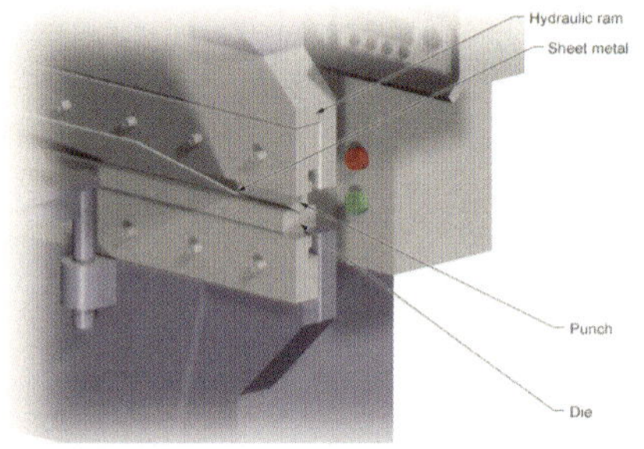


Press Brake (Closed)

While using a press brake and standard die sets, there are still a variety of techniques that can be used to bend the sheet. The most common method is known as V-bending, in which the punch and die are "V" shaped. The punch pushes the sheet into the "V" shaped groove in the V-die, causing it to bend. If the punch does not force the sheet to the bottom of the die cavity, leaving space or air underneath, it is called "air bending". As a result, the V-groove must have a sharper angle than the angle being formed in the sheet. If the punch forces the sheet to the bottom of the die cavity, it is called "bottoming". This technique allows for more control over the angle because there is less springback. However, a higher tonnage press is required. In both techniques, the width of the "V" shaped groove, or die opening, is typically 6 to 18 times the sheet thickness. This value is referred to as the die ratio and is equal to the die opening divided by the sheet thickness.

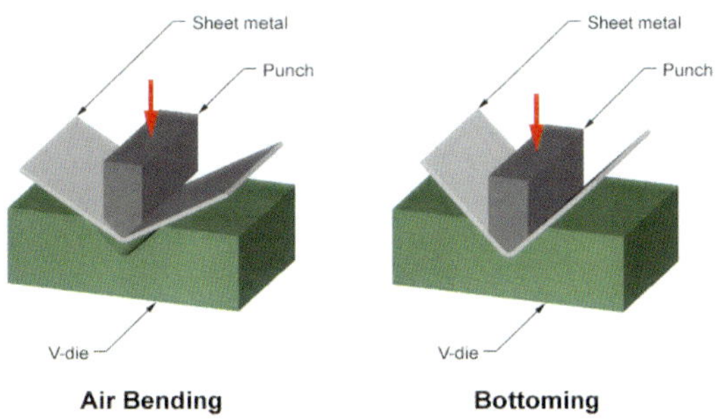


V Bending

In addition to V-bending, another common bending method is wipe bending, sometimes called edge bending. Wipe bending requires the sheet to be held against the wipe die by a pressure pad. The punch then presses against the edge of the sheet that extends beyond the die and pad. The sheet will bend against the radius of the edge of the wipe die.

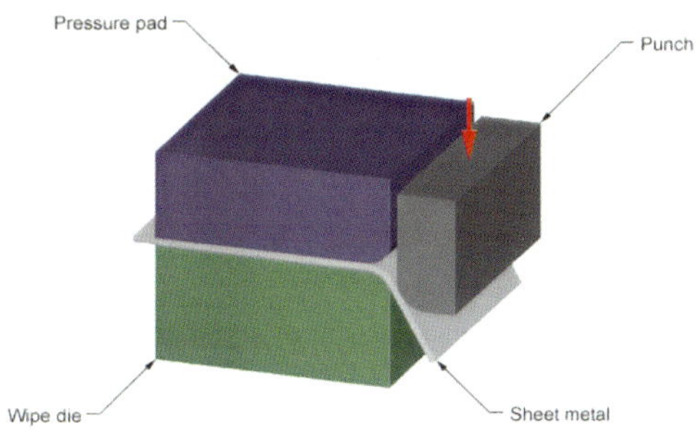


Wipe Bending

## Design Rules

Bend location - A bend should be located where enough material is present, and preferably with straight edges, for the sheet to be secured without slipping. The width of this flange should be equal to at least 4 times the sheet thickness plus the bend radius.

## Bend Radius

Use a single bend radius for all bends to eliminate additional tooling or setups

Inside bend radius should equal at least the sheet thickness

Bend direction - Bending hard metals parallel to the rolling direction of the sheet may lead to fracture. Bending perpendicular to the rolling direction is recommended.

Any features, such as holes or slots, located too close to a bend may be distorted. The distance of such features from the bend should be equal to at least 3 times the sheet thickness plus the bending radius.

In the case of manual bending, if the design allows, a slot can be cut along the bend line to reduce the manual force required.

## Roll Forming

Roll forming, sometimes spelled rollforming, is a metal forming process in which sheet metal is progressively shaped through a series of bending operations. The process is performed on a roll forming line in which the sheet metal stock is fed through a series of roll stations. Each station has a roller, referred to as a roller die, positioned on both sides of the sheet. The shape and size of the roller die may be unique to that station, or several identical roller dies may be used in different positions. The roller dies may be above and below the sheet, along the sides, at an angle, etc. As the sheet is forced through the roller dies in each roll station, it plastically deforms and bends. Each roll station performs one stage in the complete bending of the sheet to form the desired part. The roller dies are lubricated to reduce friction between the die and the sheet, thus reducing the tool wear. Also, lubricant can allow for a higher production rate, which will also depend on the material thickness, number of roll stations, and radius of each bend. The roll forming line can also include other sheet metal fabrication operations before or after the roll forming, such as punching or shearing.

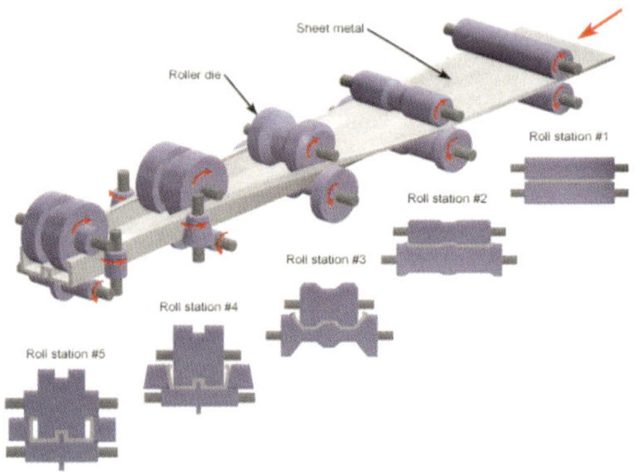


Roll Forming Line

The roll forming process can be used to form a sheet into a wide variety of cross-section profiles. An open profile is most common, but a closed tube-like shape can be created as well. Because the final form is achieved through a series of bends, the part does not require a uniform or symmetric cross-section along its length. Roll forming is used to create very long sheet metal parts with typical widths of 1-20 inches and thicknesses of 0.004-0.125 inches. However wider and thicker sheets can be formed, some up to 5 ft. wide and 0.25 inches thick. The roll forming process is capable of producing parts with tolerances as tight as ±0.005 inches. Typical roll formed parts include panels, tracks, shelving, etc. These parts are commonly used in industrial and commercial buildings for roofing, lighting, storage units, and HVAC applications.

## Spinning

Spinning, sometimes called spin forming, is a metal forming process used to form cylindrical parts by rotating a piece of sheet metal while forces are applied to one side. A sheet metal disc is rotated at high speeds while rollers press the sheet against a tool, called a mandrel, to form the shape of the desired part. Spun metal parts have a rotationally symmetric, hollow shape, such as a cylinder, cone, or hemisphere. Examples include cookware, hubcaps, satellite dishes, rocket nose cones, and musical instruments.

Spinning is typically performed on a manual or CNC lathe and requires a blank, mandrel, and roller tool. The blank is the disc-shaped piece of sheet metal that is pre-cut from sheet stock and will be formed into the part. The mandrel is a solid form of the internal shape of the part, against which the blank will be pressed. For more complex parts, such as those with reentrant surfaces, multi-piece mandrels can be used. Because the mandrel does not experience much wear in this process, it can be made from wood or plastic. However, high volume production typically utilizes a metal mandrel. The mandrel and blank are clamped together and secured between the headstock and tailstock of the lathe to be rotated at high speeds by the spindle. While the blank and mandrel rotate, force is applied to the sheet by a tool, causing the sheet to bend and form around the mandrel. The tool may make several passes to complete the shaping of the sheet. This tool is usually a roller wheel attached to a lever. Rollers are available in different diameters and thicknesses and are usually made from steel or brass. The rollers are inexpensive and experience little wear allowing for low volume production of parts.

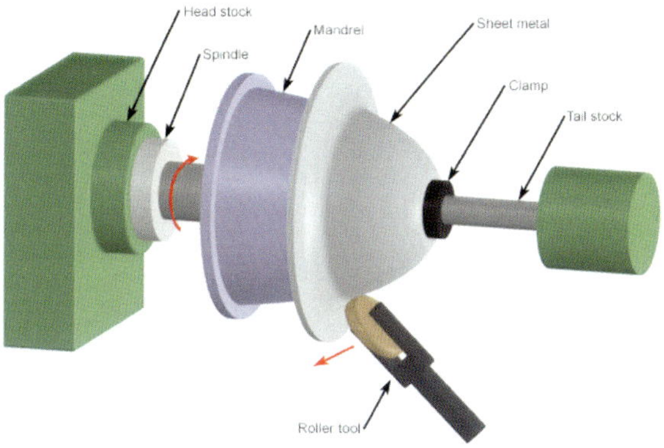


## Spinning Lathe

There are two distinct spinning methods, referred to as conventional spinning and shear spinning. In conventional spinning, the roller tool pushes against the blank until it conforms to the contour of the mandrel. The resulting spun part will have a diameter smaller than the blank, but will maintain a constant thickness. In shear spinning, the roller not only bends the blank against the mandrel, it also applies a downward force while it moves, stretching the material over the mandrel. By doing so, the outer diameter of the spun part will remain equal to the original blank diameter, but the thickness of the part walls will be thinner.

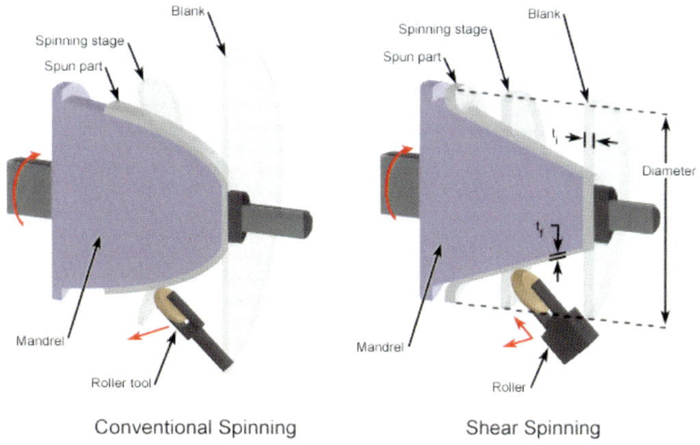


Conventional Spinning vs. Shear Spinning

## Deep Drawing

Deep drawing is a metal forming process in which sheet metal is stretched into the desired part shape. A tool pushes downward on the sheet metal, forcing it into a die cavity in the shape of the desired part. The tensile forces applied to the sheet cause it to plastically deform into a cup-shaped part. Deep drawn parts are characterized by a depth equal to more than half of the diameter of the part. These parts can have a variety of cross sections with straight, tapered, or even curved walls, but cylindrical or rectangular parts are most common. Deep drawing is most effective with ductile metals, such as aluminum, brass, copper, and mild steel. Examples of parts formed with deep drawing include automotive bodies and fuel tanks, cans, cups, kitchen sinks, and pots and pans.

The deep drawing process requires a blank, blank holder, punch, and die. The blank is a piece of sheet metal, typically a disc or rectangle, which is pre-cut from stock material and will be formed into the part. The blank is clamped down by the blank holder over the die, which has a cavity in the external shape of the part. A tool called a punch moves downward into the blank and draws, or stretches, the material into the die cavity. The movement of the punch is usually hydraulically powered to apply enough force to the blank. Both the die and punch experience wear from the forces applied to the sheet metal and are therefore made from tool steel or carbon steel. The process of drawing the part sometimes occurs in a series of operations, called draw reductions. In each step, a punch forces the part into a different die, stretching the part to a greater depth each time. After a part is completely drawn, the punch and blank holder can be raised and the part removed from the die. The portion of the sheet metal that was clamped under the blank holder may form a flange around the part that can be trimmed off.

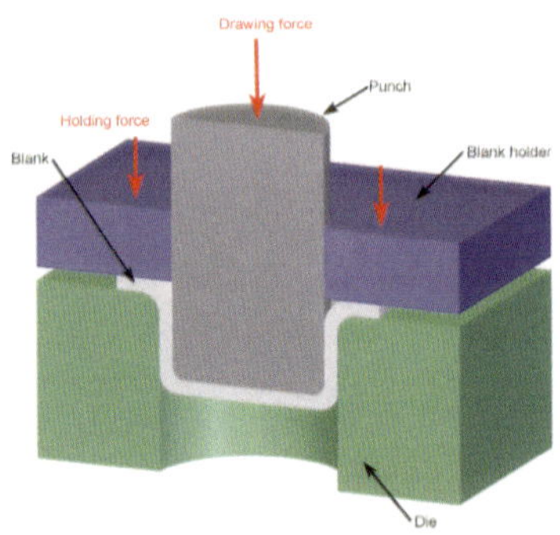


Deep Drawing

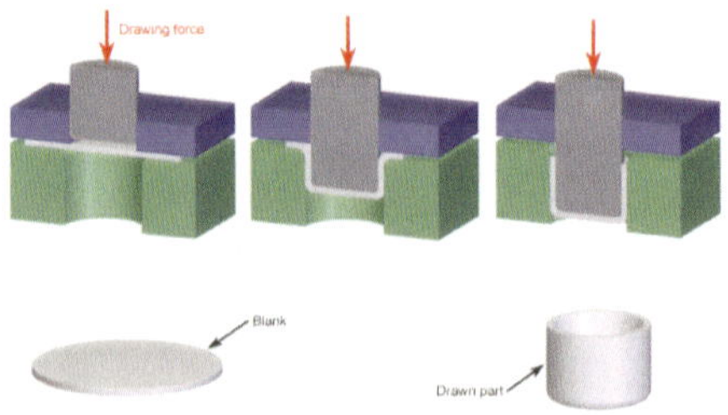


Deep Drawing Sequence

## Stretch Forming

Stretch forming is a metal forming process in which a piece of sheet metal is stretched and bent simultaneously over a die in order to form large contoured parts. Stretch forming is performed on a stretch press, in which a piece of sheet metal is securely gripped along its edges by gripping jaws. The gripping jaws are each attached to a carriage that is pulled by pneumatic or hydraulic force to stretch the sheet. The tooling used in this process is a stretch form block, called a form die, which is a solid contoured piece against which the sheet metal will be pressed. The most common stretch presses are oriented vertically, in which the form die rests on a press table that can be raised into the sheet by a hydraulic ram. As the form die is driven into the sheet, which is gripped tightly at its edges, the tensile forces increase and the sheet plastically deforms into a new shape. Horizontal stretch presses mount the form die sideways on a stationary press table, while the gripping jaws pull the sheet horizontally around the form die.

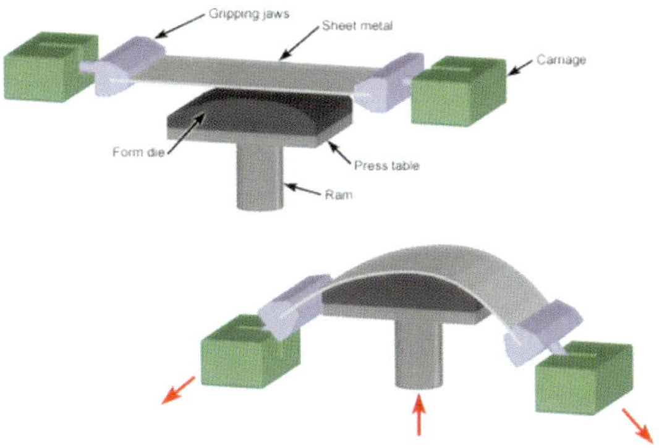


Stretch Forming

Stretch formed parts are typically large and possess large radius bends. The shapes that can be produced vary from a simple curved surface to complex non-uniform cross sections. Stretch forming is capable of shaping parts with very high accuracy and smooth surfaces. Ductile materials are preferable, the most commonly used being aluminum, steel, and titanium. Typical stretch formed parts are large curved panels such as door panels in cars or wing panels on aircraft. Other stretch formed parts can be found in window frames and enclosures.

## Forging Process

Forging technology occupies a very important place among all the manufacturing processes as it produces parts with excellent properties and with minimal wastage. Through a series of operations, metals with very simple geometry are physically deformed into products of complex configuration.

In the forging process the metal is heated up by applying compressive force and given shape by plastic deformation. The compressive force is applied by hammer blows using a power hammer or a press. Powered by compressed air, hydraulics electricity or steam. The weight of the hammer can be 500 pounds to thousands of pounds. Forging has the capacity to refine the grain structure and improve the physical properties of the metal.

Forging products are consistent, without the defects of porosity, inclusion or voids, finishing operations like machining and coating operations like painting or plating can also be done very easily.

Typically a forged metal results in the following:

- Drawing out of the metal: Increased length decreased cross-section.
- Upsetting the metal: Decrease in length decreased cross-section.
- Change in length; change in cross-section: Resulting in favourable grain flow for strong parts.

## Methods of Forging

### Impression Die Forgings

Impression die forging accounts for the majority of forging production. In this process, two dies are brought together. The workpiece is then subjected to plastic deformation till its enlarged sides touch the die side walls. Materials then begins to flow outside the die impression forming flash. As the flash cools down and increase its resistance to deformation, in effect it becomes part of the tool itself. This creates pressure in the bulk of the workpiece and in the process helping the material to flow into the unfilled impression.

Precision die forgings that is produced on a horizontal forging machine (upsetter) is known as upsetting. In upsetting, stock is held between a fixed and moving die while a horizontal ram provides the pressure to forge the stock. After each ram stroke, the multiple-impression dies opens to permit transfer of stock from one cavity to another.

Closed die forging is another form of impression die forging. It does not depend on flash formation to achieve complete filling of the die. Material is deformed in a cavity that allows little or no escape of excess material, thus placing greater demands on die design.

### Net Shape

As, it's a common occurrence that forging results in wastage of material in the form of material flash and machining operations. This can reach 70% for gear blanks and even around 90% in case of aircraft structural parts. Net-shape and near-net-shape processes helps in minimizing this enormous waste by making precision dies and making parts with very little draft angle (less than $1^{o}$).

## Limitation

This process is expensive in terms of tooling requirements and the capital expenditure involved. It can be only suited for processes that are very wasteful and the material savings made will pay for the rise in tooling costs.

## Open Die Forgings

Open die forgings are done with repeated blows in an open die, with the operator manipulating the workpiece in the die. The process is called Open Die Forgings as the metal does not remain confined laterally by impression dies during the forging process. Generally flat dies are used for open die forgings. Round swaging dies and V dies can also be used together or with a flat die. Generally heavy parts like shafts are forged. Part weights can range from 5 to 5,00,000 lbs.

Open die forgings can forge all ferrous and non-ferrous alloys, corrosion-resistant refractory alloys and materials like age-hardening super alloys

## Operating Open Die Forgings

- First the ingot or billet has to be lengthened either by reducing the cross-section or by drawing out.
- Enlarging the diameter of the ingot or billet by upsetting or reducing the length.
- Sometimes to forge a rough contoured ring a combination of upsetting, drawing out and piercing method is combined with forging.

The forging workpiece is continuously hammered or pressed, and manipulated repeatedly between the dies until it reaches final forged dimensions. The process is not exactly precise and considerable workpiece stock allowances are kept to accommodate forging irregularities. The forged part is then machined (both rough and finish) to final dimensions. The metal to be forged is always hot worked above their recrystallization temperatures. It is done as the repeated change in the work-piece positioning makes its temperature fall below the recrystallization temperatures. It has to be re-heated before further forging takes place. For e.g a 24 ft long and 2 ft diameter shaft requires four to six heats before the forged dimension is achieved.

A sub process of open die forgings is the compression between flat dies, or Upsetting process. Whereby an oblong workpiece is placed at the end of a lower die whose height is reduced by the downward movement of the top die. The friction between the end faces of the workpiece and dies prevents the free lateral spread of the metal. Thus a barrel shape of the metal is produced. The barrel is resistant to deformation as contact with the cool die surface chills the end faces of the metal.

## Press Forgings

In this process a slow continuous pressure is applied to the area to be forged. The pressure extends deep into the material. Press forging can be either cold or hot. Cold press forging is generally used on thin, annealed materials, while a hot press forging is done on works such as locomotive armor plating and heavy machinery. Cost wise, press forging is more economical than hammer forging (except when low production is required) and closer tolerances can be achieved. A greater quantity of the work done is passed on to the workpiece. This way it differs from the hammer forging operation, where most of the work is absorbed by the machine and foundation. This production process is ideally suited for larger forgings, as there is no limitation in the size of the machine.

## Roll Forgings

In the process of roll forging, a bar stock, round or flat is placed between die rollers which reduces the cross-section and increases the length to form parts such as axles, leaf springs etc. Roll forgings has a wide spread use. As it can easily be made continuous by the use of multiple roller and dies. Roll forging can also be said to be a type of draw forging process as the workpiece is gradually drawn out into the requisite shape.

## Application of Roll Forgings

Propeller blades, tapered axle shaft shapes, continuous uniform shapes.

## Swaging

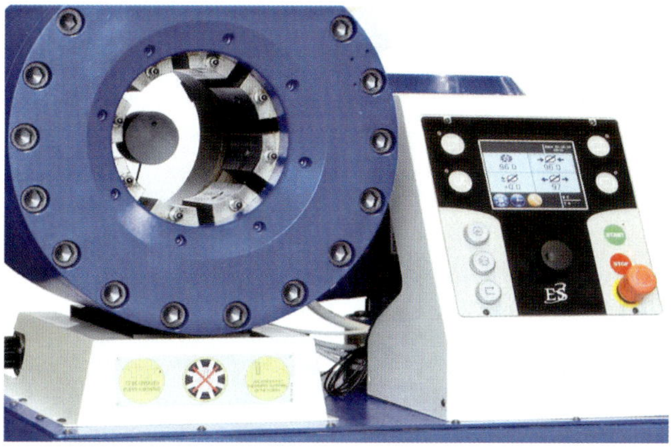

This method is related to the open die forging process where a tube or rod is forced inside a die and the diameter gets reduced as the cylindrical object is fed. The die hammer is rotated to apply multiple blows at the diameter which can go up to 2000 per minute. This causes the metal to flow inward making the outer diameter of the tube or rod to take the shape of the die.

## Upset Forgings

It's the production process under which high pressure deforms plastically metal into high strength components. It is also called hot heading process. Using this process cross-sectional size of a bar can be increased, either at ends or at some point along the length. Specially designed upsetting machines using closed dies are used to control size and shape. They can be made in various sizes. The mechanical press producing these components operates horizontally. The dies are split up allowing material to reach beyond the mechanical press. The forming force is supplied by a third die attached to the header. The process when done cold is called cold heading.

## Advantages of Upset Forgings

- Greater Strength:As the material grain flow is oriented to the component's shape, compared to parts machined from bar stock, upset forgings offer greater strength.
- Piercing and Trimming Operations:This contribute towards component weight reduction and eliminate machining operations.
- Economical Production:As Upset forgings require fewer operations, have lower scrap costs, that can result in more cost efficient production.

## Application of Upset Forgings

Artillery shells, cluster gear blanks, heads of bolts, valves, single and cylinders for radial engines are examples of parts made by upset forging.

## Applications of Forging Process

Wide variety of uses in different kinds of Industries:

Automobile Industry: Wheel spindles, kingpins, axle beams and shafts, torsion bars, ball studs, idler arms and steering arm.

Agro-Industries: Engine and transmission components, levers, gears, shafts and spindles to tie-rod ends, spike harrow teeth and cultivator shafts.

Aerospace: Bulkheads, hinges, wing roots, engine mounts, brackets, beams, shafts, landing gear cylinders and struts, wheels, brake carriers and discs and arresting hooks, blades, buckets couplings etc.

Hand Tools: Sledges, pliers, hammers, wrenches and garden tools, as well as wire-rope clips, sockets, hooks, turnbuckles and eye bolts are common examples.

Industrial Equipment: Connecting rods, blanks, blocks, cylinders, discs, elbows, rings, T's, shafts and sleeves.

## Tips for Selecting the Right Forging Technology

- Technology to remain competitive must come out with cost effective alternatives. That is the reason computer aided techniques like CAD, CAM, CAE and Finite Element Analysis (FEA) based computer simulation, are used to selecting the right forging process.
- Understanding the forged material's flow behaviour under processing conditions.

- Knowledge of the die geometry and materials.
- Environmental considerations.
- Evaluating the mechanics of deformation process-stress and strain.
- Friction and Lubricating process.
- Nature of the Forging equipment.

## Rolling

Rolling is a fabricating process in which the metal, plastic, paper, glass, etc. is passed through a pair (or pairs) of rolls. There are two types of rolling process, flat and profile rolling. In flat rolling the final shape of the product is either classed as sheet (typically thickness less than 3 mm, also called "strip") or plate (typically thickness more than 3 mm). In profile rolling the final product may be a round rod or other shaped bar, such as a structural section (beam, channel, joist etc). Rolling is also classified according to the temperature of the metal rolled. If the temperature of the metal is above its recrystallization temperature, then the process is termed as hot rolling. If the temperature of the metal is below its recrystallization temperature, the process is termed as cold rolling. Another process also termed as 'hot bending' is induction bending, whereby the section is heated in small sections and dragged into a required radius.

Heavy plates tend to be formed using a press process, which is termed forming, rather than rolling.

### Hot Rolling

Hot rolling is a hot working metalworking process where large pieces of metal, such as slabs or billets, are heated above their recrystallization temperature and then deformed between rollers to form thinner cross sections. Hot rolling produces thinner cross sections than cold rolling processes with the same number of stages. Hot rolling, due to recrystallization, will reduce the average grain size of a metal while maintaining an equiaxed microstructure where as cold rolling will produce a hardened microstructure.

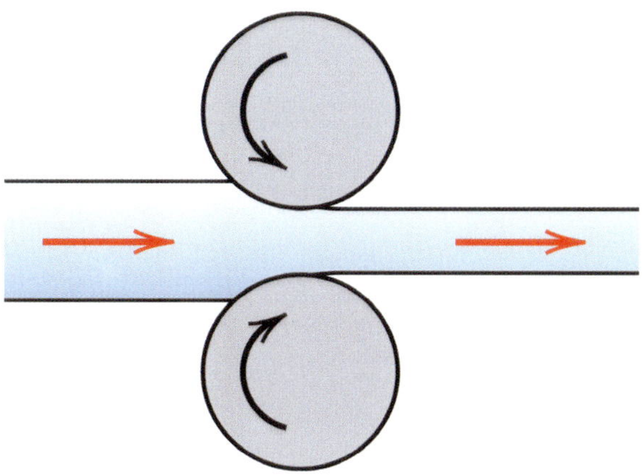

Hot/Cold Rolling

## Hot Rolling Process

A slab or billet is passed or deformed between a set of work rolls and the temperature of the metal is generally above its recrystallization temperature, as opposed to cold rolling, which takes place below this temperature. Hot rolling permits large deformations of the metal to be achieved with a low number of rolling cycles. As the rolling process breaks up the grains, they recrystallize maintaining an equiaxed structure and preventing the metal from hardening. Hot rolled material typically does not require annealing and the high temperature will prevent residual stress from accumulating in the material resulting better dimensional stability than cold worked materials.

Hot rolling is primarily concerned with manipulating material shape and geometry rather than mechanical properties. This is achieved by heating a component or material to its upper critical temperature and then applying controlled load which forms the material to a desired specification or size.

## Hot Rolling Applications

Hot rolling is used mainly to produce sheet metal or simple cross sections such as rail road bars from billets.

Mechanical properties of the material in its final 'as-rolled' form are a function of:

- material chemistry,
- reheat temperature,
- rate of temperature decrease during deformation,
- rate of deformation,
- heat of deformation,
- total reduction,
- recovery time,
- recrystallisation time, and
- subsequent rate of cooling after deformation.

## Types of Hot Rolling Mills

Prior to continuous casting technology, ingots were rolled to approximately 200 millimetres (7.9 in) thick in a slab or bloom mill. Blooms have a nominal square cross section, whereas slabs are rectangular in cross section.

Slabs are the feed material for hot strip mills or plate mills and blooms are rolled to billets in a billet mill or large sections in a structural mill.

The output from a strip mill is coiled and, subsequently, used as the feed for a cold rolling mill or used directly by fabricators. Billets, for re-rolling, are subsequently rolled in either a merchant, bar or rod mill.

Merchant or bar mills produce a variety of shaped products such as angles, channels, beams, rounds (long or coiled) and hexagons. Rounds less than 16 millimetres (0.63 in) in diameter are more efficiently rolled from billet in a rod mill.

## Cold Rolling

Rolling mill for cold rolling metal sheet like this piece of brass sheet.

Cold rolling is a metalworking process in which metal is deformed by passing it through rollers at a temperature below its recrystallization temperature. Cold rolling increases the yield strength and hardness of a metal by introducing defects into the metal's crystal structure. These defects prevent further slip and can reduce the grain size of the metal, resulting in Hall-Petch hardening.

Cold rolling is most often used to decrease the thickness of plate and sheet metal.

## Physical Metallurgy of Cold Rolling

Cold rolling is a method of cold working a metal. When a metal is cold worked, microscopic defects are nucleated throughout the deformed area. These defects can be either point defects (a vacancy on the crystal lattice) or a line defect (an extra half plane of atoms jammed in a crystal). As defects accumulate through deformation, it becomes increasingly more difficult for slip, or the movement of defects, to occur. This results in a hardening of the metal.

If enough grains split apart, a grain may split into two or more grains in order to minimize the strain energy of the system. When large grains split into smaller grains, the alloy hardens as a result of the Hall-Petch relationship. If cold work is continued, the hardened metal may fracture.

During cold rolling, metal absorbs a great deal of energy. Some of this energy is used to nucleate and move defects (and subsequently deform the metal). The remainder of the energy is released as heat.

While cold rolling increases the hardness and strength of a metal, it also results in a large decrease in ductility. Thus metals strengthened by cold rolling are more sensitive to the presence of cracks and are prone to brittle fracture.

A metal that has been hardened by cold rolling can be softened by annealing. Annealing will relieve stresses, allow grain growth, and restore the original properties of the alloy. Ductility is also restored by annealing. Thus, after annealing, the metal may be further cold rolled without fracturing.

## Degree of Cold Work

Cold rolled metal is given a rating based on the degree it was cold worked. "Skin-rolled" metal undergoes the least rolling, being compressed only 0.5-1% to harden the surface of the metal and make it more easily workable for later processes. Higher ratings are "quarter hard," "half hard" and "full hard"; in the last of these, the thickness of the metal is reduced by 50%.

## Cold Rolling as a Manufacturing Process

Cold rolling is a common manufacturing process. It is often used to form sheet metal. Beverage cans are closed by rolling, and steel food cans are strengthened by rolling ribs into their sides. Rolling mills are commonly used to precisely reduce the thickness of strip and sheet metals.

## Foil Rolling

Foil rolling is a continuous deformation process compressing metal between a pair of rollers called work rolls. Another term for this process is called plate roll bending. This is defined as a cold forming process where plate and steel metal is formed into cylindrical shapes by a combination of three rolls arranged in a pyramid formation. Two of the rolls are power driven, in a fixed position, and the third is adjustable to suit the desired bend radius and workpiece thickness.

Foil is produced for several applications:

- Thermal insulation for the construction industry
- Fin stock for air conditioners
- Electrical coils for transformers
- Capacitors for radios and televisions
- Insulation for storage tanks
- Decorative products
- Containers and packaging

Foil stock is reduced in thickness by a rolling mill, where the material is passed several times through metal work rolls. As the sheets of metal pass through the rolls, they are squeezed thinner and extruded through the gap between the rolls. The work rolls are paired with heavier rolls called backup rolls, which apply pressure to help maintain the stability of the work rolls. The work and backup rolls rotate in opposite directions. As the foil sheets come through the rollers, they are trimmed and slitted with circular or razor-like knives installed on the rolling mill. Trimming refers to the edges of the foil, while slitting involves cutting it into several sheets.

Aluminium alloys are most commonly produced in the foil rolling process because the raw materials necessary for its manufacture are abundant. Aluminium foil is inexpensive, durable, non-toxic, and greaseproof. Iron, silicon, and manganese are all major alloying elements. Sheet metals with a thickness below 200 micrometers are considered foils (Some foils may be as thin as 6.3 micrometers).

## Plate Roll Bending

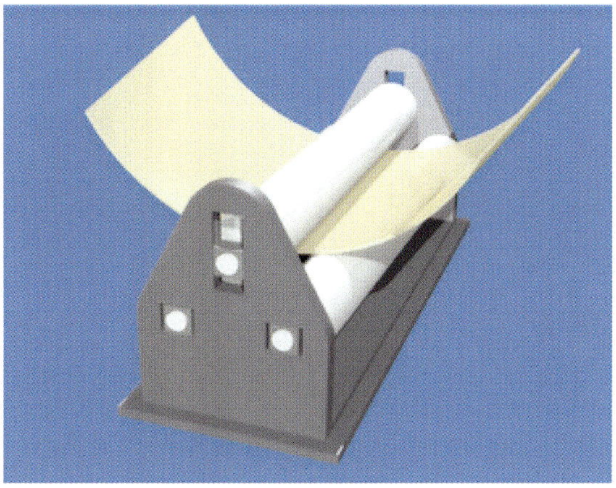

Profile rolling (to manufacture a cone)

Plate Roll Bending is a cold forming process. Plate or steel metal is formed into cylindrical shapes by a combination of three rolls arranged in a pyramid formation. Two of the rolls are power driven, in a fixed position, and the third is adjustable to suit the desired bend radius and workpiece thickness.

## Workpiece Geometry

Shapes produced range from simple cylindrically shaped parts to more complex parts, such as conical and flattened from 1/16in to 6in, and lengths of up to 20in, or more, can be formed by this method.

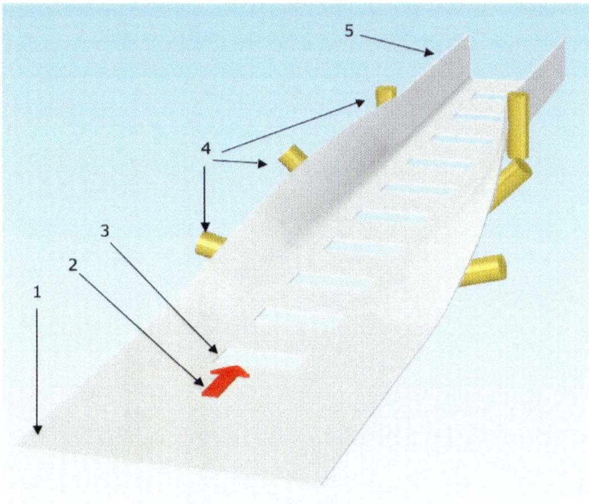


Roll Forming

## Progressive Roll Forming

Roll forming, also spelled rollforming, is a continuous bending operation in which a long strip of metal (typically coiled steel) is passed through consecutive sets of rolls, or stands, each performing only an incremental part of the bend, until the desired cross-section profile is obtained. Roll forming is ideal for producing parts with long lengths or in large quantities.

### Geometric Possibilities

The geometric possibilities can be very broad and even include enclosed shapes so long as it is the same cross-section throughout. Typical sheeting thicknesses range from 0.025in. to 0.25in., but they can exceed that. Length is almost unaffected by the rolling process. The part widths typically aren't smaller than 1in. however they can exceed 20in.

## Extrusion

Extrusion is the process by which long straight metal parts can be produced. The cross-sections that can be produced vary from solid round, rectangular, to L shapes, T shapes. Tubes and many other different types. Extrusion is done by squeezing metal in a closed cavity through a tool, known as a die using either a mechanical or hydraulic press.

Extrusion produces compressive and shear forces in the stock. No tensile is produced, which makes high deformation possible without tearing the metal. The cavity in which the raw material is contained is lined with a wear resistant material. This can withstand the high radial loads that are created when the material is pushed the die.

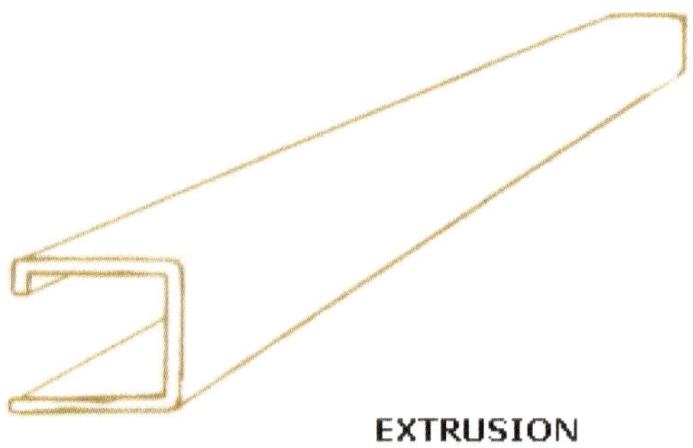

EXTRUSION

### Features of Extrusion Process

- Cost effective: Minimizes the need for secondary machining process.
- Surface finish: For steel is 3 μm; (125 μ in), for Aluminum and Magnesium -0.8 μm (30 μ in).
- Cross-section: Wide variety of cross-sections can be made.
- Minimum thickness: For steel 3 mm (0.120 in), for Aluminum and Magnesium 1mm (0.040 in).
- Minimum cross section: For steel 250 mm (0.4 in) for steel.
- Corner and fillet radii: 0.4 mm (0.015 in) for Aluminum and Magnesium, for steel the minimum corner radius is 0.8mm (0.030 in) and 4 mm (0.120 in) fillet radius.

## An Example of Hot Extrusion Process using Aluminium Alloy

The alloyed press bars are cut into smaller pieces and heated up in an induction furnace to 450 to 500°C. The bar is then pressed with very high force using speeds between 5 to 50 m/min through a hollowed tool. As a result a profile is formed. The length of the profile ranges between 25 to 45 m. Immediately after the pressing operation, profile is cooled with air or water.

The profile is straightened and internal stresses released by stretching it in a pulling machine just after cooling. The profile is then cut into required lengths. Finally ageing gives the strength of the material. Which can be done by natural ageing at normal temperature or artificial ageing done at elevated temperature of 170 to 185°C.

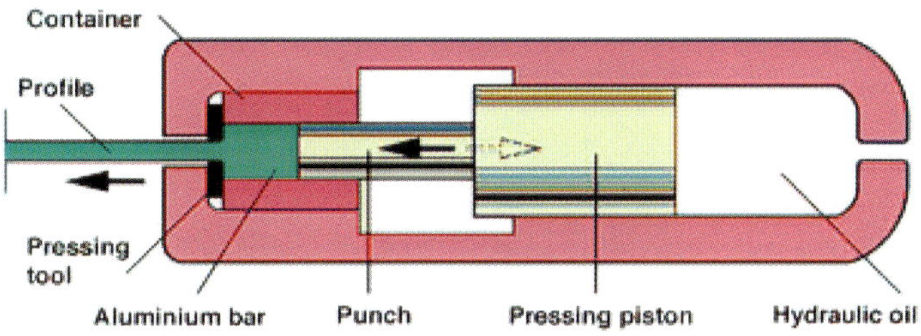


Extrusion Process

Application of Extrusion Process: Trim parts as used in automobile and construction equipment, railings, window frame members, structural parts etc.

## Types of Extrusion

### Hot Extrusion

Generally done at fairly high temperatures, approximately at 50 to 75 % of the melting point of the metal. The pressures range from 35-700 MPa (5076 - 101,525 psi). To cool down the high temperatures and pressures and its adverse effect on the die life as well as other components, good lubrication is a must. Oil graphite and glass powder is preferred as lubricants.

### Cold Extrusion

Cold extrusion takes place at room temperature or slightly elevated temperatures. This process is useful for withstanding the stresses created by extrusion.

The advantages of cold extrusion are:

- No oxidation process.
- Good mechanical properties provided the temperatures created are below the re-crystallization temperature. Good surface finish

## Types of Extrusion

Cold Extrusion: Cold extrusion is the process done at room temperature or slightly elevated temperatures. This process can be used for most materials-subject to designing robust enough tooling that can withstand the stresses created by extrusion. Examples of the metals that can be extruded are lead, tin, aluminum alloys, copper, titanium, molybdenum, vanadium, steel. Examples of parts that are cold extruded are collapsible tubes, aluminum cans, cylinders, gear blanks. The advantages of cold extrusion are:

| | |
|---|---|
| • | No oxidation takes place. |
| • | Good mechanical properties due to severe cold working as long as the temperatures created are below the re-crystallization temperature. |
| • | Good surface finish with the use of proper lubricants. |

## Application of Cold Extrusion

Examples of the metals that can be extruded are copper, lead, tin, aluminum alloys, titanium, molybdenum, vanadium, steel. Which are used to make parts like collapsible tubes, gear blanks, aluminum cans, cylinders etc.

In automobile sector they have found wide applications in Injection technology; Engine control; Fuel supply; Automatic transmissions Seat technology; Safety systems (restraint systems).

Typical parts produced by extrusions are trim parts used in automotive and construction applications, window frame members, railings, aircraft structural parts.

## Application of Hot Extrusion

Aluminium, copper with their alloys are successfully used to manufacture products using hot extrusion process. Electrical wires, bars and tubes are some of the items produced.

# Swaging

In this process, the diameter of a rod or a tube is reduced by forcing it into a confining die. A set of reciprocation dies provides radial blows to cause the metal to flow inward and acquire the form of the die cavity. The die movements may be of in – and – out type or rotary. The latter type is obtained with the help of a set of rollers in a cage, in a similar action as in a roller bearing. The workpiece is held stationary and the dies rotate, the dies strike the workpiece at a rate as high as 10 - 20 strokes per second.

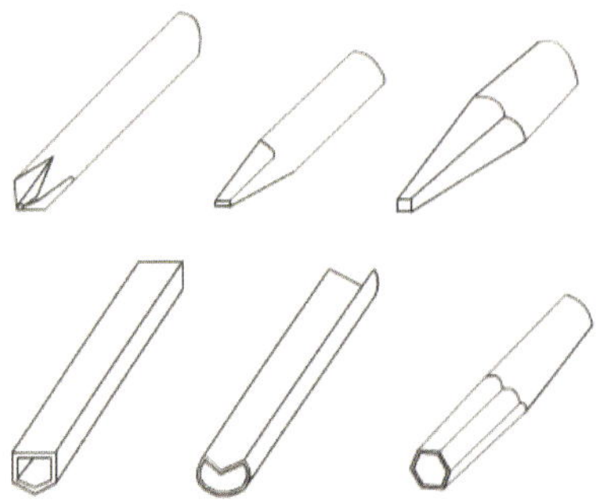

Figure: Typical parts made by swaging.

Screwdriver blades and soldering iron tips are typical examples of swaged products. Figure shows these and other products made by swaging.

In tube swaging, the tube thickness and / or internal dia of tube can be controlled with the use of internal mandrels. For small – diameter tubing, a thin rod can be used as a mandrel; even internally shaped tubes can be swaged by using shaped mandrels. Figure shows the process.

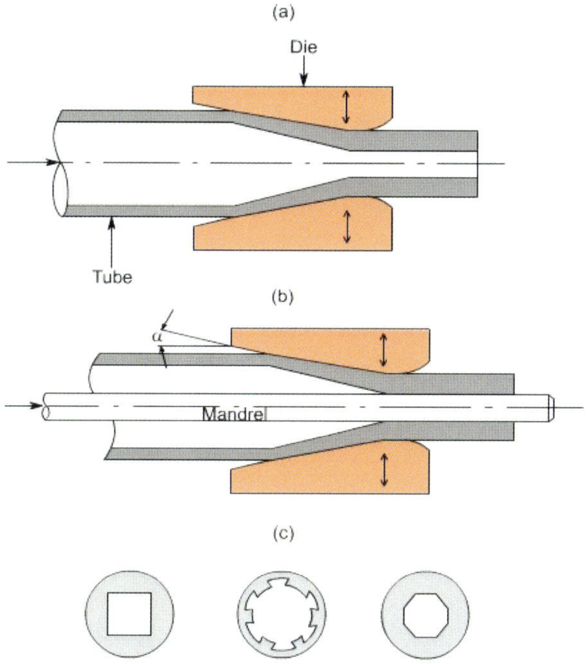


Figure: (a) Swaging of tubes without a mandrel. Wall thickness is more in the die gap.
(b) Swaging with a mandrel. The final wall thickness of the tube depends on the mandrel diameter.
(c) Examples of cross-sections of tubes produced by swaging on shaped mandrels.

The process is quite versatile. The maximum diameter of work piece that can be swaged is limited to about 150 mm; work pieces as small as 0.5 mm diameter have been swaged. The production rate can be as high as 30 parts per minute depending upon the complexity of the part shape and the part handling means adopted.

The parts produced by swaging have tolerance in the range ± 0.05 mm to ± 0.5 mm and improved mechanical properties. Use of lubricants helps in obtaining better work surface finish and longer die life. Materials, such as tungsten and molybdenum are generally swaged at elevated temperatures as they have low ductility at room temperature. Hot swaging is also used to form long or steep tapers, and for large reductions.

Swaging is a noisy operation. The level of noise can be, however, reduced by proper mounting of the machine or by the use of enclosure.

## Wire Drawing

Wire drawing is primarily the same as bar drawing except that it involves smaller – diameter material that can be coiled. It is generally performed as a continuous operation on draw bench like the one shown in figure.

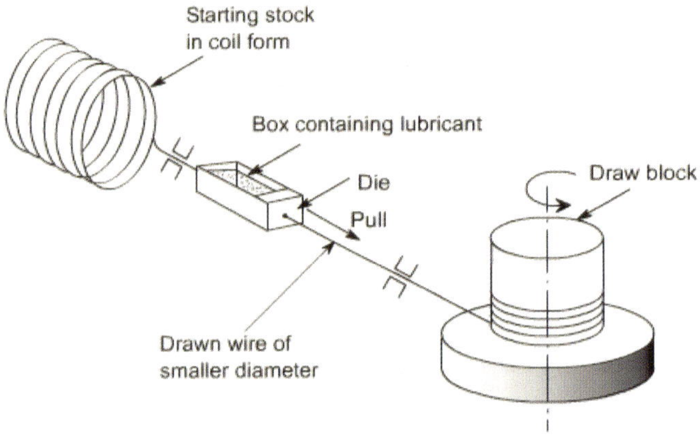


Figure: Wire drawing on a continuous draw block. The rotating draw block provides a continuous pull on the incoming wire.

Large coil of hot rolled material of nearly 10 mm diameter is taken and subjected to preparation treatment before the actual drawing process. The preparation treatment for steel wire consists of :

- Cleaning. This may be done by acid pickling, rinsing, and drying. Or, it may be done by mechanical flexing.
- Neutralization. Any remaining acid on the raw material is neutralized by immersing it in a lime bath. The corrosion protected material is also given a thin layer of lubricant.

To begin the drawing process, one end of coil is reduced in cross section upto some length and fed through the drawing die, and gripped. A wire drawing die is generally made of tungsten carbide and has the configuration shown in Figure for drawing very fine wire, diamond die is preferred.

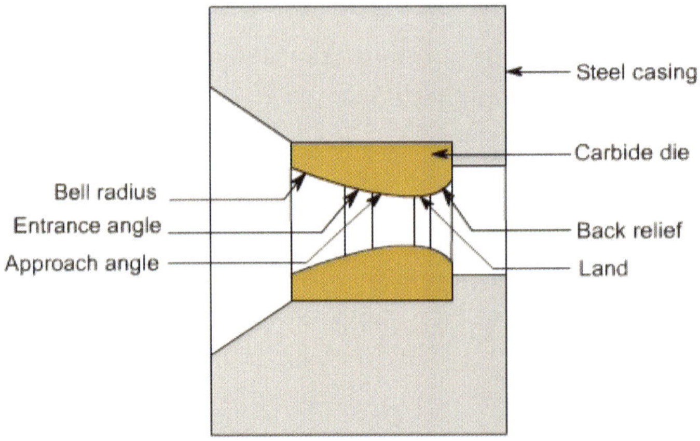


Figure: Cross section through a typical carbide wire drawing die.

Small diameter wire is generally drawn on tandom machines which consists of a series of dies, each held in a water – cooled die block. Each die reduces the cross section by a small amount so as to avoid excessive strain in the wire. Intermediate annealing of material between different states of wire may also be done, if required.

## Wire Drawing Terms

$$\text{Draft} = D_o - D_f$$

$$\%\ \text{reduction in area} \ \frac{A_o - A_f}{A_f} \times 100 = \frac{D_o^2 - D_f^2}{D_f^2} \times 100$$

$$\%\ \text{elongation} = \frac{L_f - L_o}{L_o}$$

Where Do , Df , Lo and Lf are the original and final diameter and length. Ao and Af are original and final cross sectional area.

For a single cold – drawing pass, the percent area reduction that can be done depends upon many factors. These include the type of material, its size, initial metallurgical condition, the final size and mechanical properties desired, die design and lubrication efficiency. The percent of area reduction per pass can range from near zero to 50%.

## Die Pull

The force required to pull the stock through the die (under frictionless conditions) can be computed as follows.

$$F = Y_{avg} .A_f \ .\text{In}\left(\frac{A_o}{A_f}\right)$$

Where F = die pull, i.e. the force required to pull the stock through the die

$Y_{avg}$ = average true stress of the material in the die gap

$A_o$, $A_f$ = original and final areas of cross section of material.

Alternatively, the following expression can be used:

$$F = c\,\sigma_t\,(A_o - A_f)$$

where c is a constant whose value is in the range 1.5 to 3.0 depending upon the % area reduction, (lower value for higher % reduction), and st is tensile strength of material before drawing.

The pull force determines the machine capacity needed.

## Tube Drawing

The diameter and wall thickness of tubes that have been produced by extrusion or other processes can be reduced by tube drawing process. The process of tube drawing is similar to wire or rod drawing except that it usually requires a mandrel of the requisite diameter to form the internal hole.

Tubes as large as 0.3 m in diameter can be drawn.

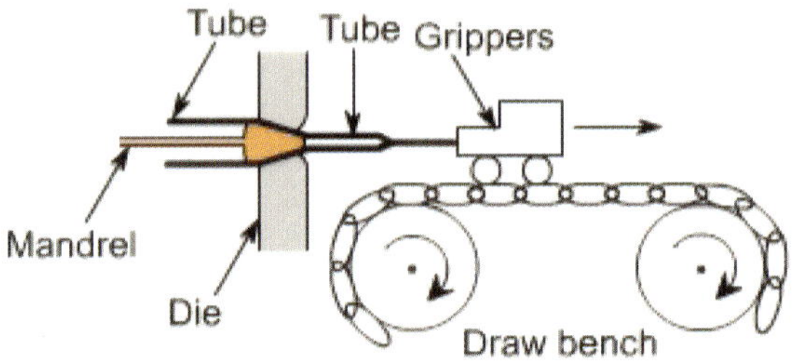


## Drawing Equipment

Drawing equipment can be of several designs. These designs can be classified into two basic types; draw bench, and bull block. A draw bench uses a single die and the pulling force is supplied by a chain drive or by hydraulic means. Draw bench is used for single length drawing of rod or tube with diameter greater than 20mm. Length can be as much as 30 m. The drawing speed attainable on a draw bench ranges from 5 m/min to 50 m/min. Draw benches are available having capacities to provide pull force of upto 1 MN.

Bull block or rotating drum is used for drawing rods or wires of very long length.

# Explosive Forming

Explosive forming is a process in which forces produced by an explosion are used to shape a workpiece. The explosive energy replaces the male part of the tool that is used in regular forming processes.

## Process Description

A die is manufactured according to the desired shape of the product. Dies can be made from a number of materials including concrete, ductile iron, fibreglass and kirksite or epoxy and concrete.

Forming takes place inside a water tank or "explosion tank" where the die is placed. The workpiece is situated directly above the die and explosive material placed above that. The explosive material is typically Trinitro toluene (TNT), a mixture of acetylene and oxygen or Nitroglycerine.

Quite simply, when the explosive material is detonated, the resulting shock wave through the transfer medium (the water) forces the workpiece material into the die, thus assuming its shape

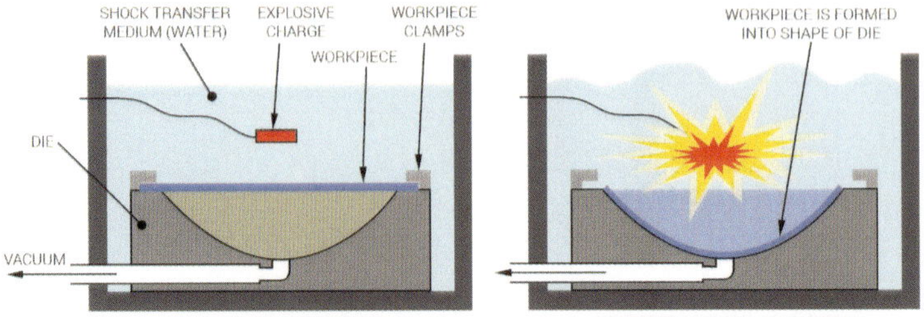


Explosive forming is an ideal process for producing low volumes of very large metal part with

low tooling costs. Sometimes called "explosive hydroforming", a workpiece and explosive is placed in a tank of water. Detonation of the explosive forces the workpiece to assume the shape of the tool.

## Types

There are two techniques of high–explosive forming: stand–off technique and the contact technique.

Standoff Technique: The sheet metal work piece blank is clamped over a die and the assembly is lowered into a tank filled with water. The air in the die is pumped out. The explosive charge is placed at some predetermined distance from the work piece. On detonation of the explosive, a pressure pulse of very high intensity is produced. A gas bubble is also produced which expands spherically and then collapses. When the pressure pulse impinges against the work piece, the metal is deformed into the die with as high velocity as 120 m/s.

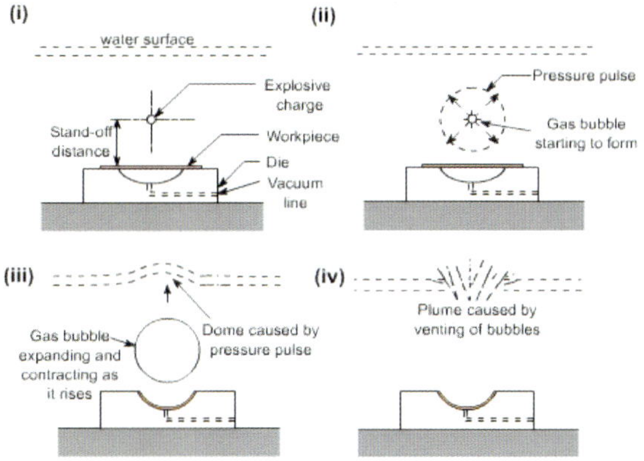


Figure: Sequeuce of underwater explosive forming operations.(i) explosive charge is set in position (ii) pressure pulse and gas bubble are formed as the detonation of charge occurs, (iii) workpiece is deformed, and (iv) gas bubbles vent at the surface of water.

The use of water as the energy transfer medium ensures a uniform transmission of energy and muffles the sound of the explosive blast. The process is versatile – a large variety of shapes can be formed, there is virtually no limit to the size of the work piece, and it is suitable for low – quantity production as well.

The process has been successfully used to form steel plates 25 mm thick x 4 m diameter and to bulge steel tubes as thick as 25 mm.

Contact Technique: The explosive charge in the form of cartridge is held in direct contact with the work piece while the detonation is initiated. The detonation builds up extremely high pressures (upto 30,000MPa) on the surface of the work piece resulting in metal deformation, and possible fracture. The process is used often for bulging tubes, as shown in figure.

## Applications

Some of the applications of explosive forming include:

- Sheet metal panels

- Tubing
- Housings
- Jet engine parts
- Missile nose cones
- Ducts

## Materials for Explosive Forming

- Both ferrous and nonferrous metals including steel, aluminum, magnesium, and their alloys.
- Some metal matrix composites like aluminum matrix, copper matrix and lead matrix composites.

## Process Variations

- Explosives can be placed at a fixed distance from the workpiece. After detonation the explosive forces travel through the intervening medium to reach the work piece. This method is called the standoff method.
- Alternatively, the explosive can be placed directly on the work piece. Upon detonation, explosive forces hit the work piece directly. This is called the contact method.

## Design Considerations

- Section thickness at any point should be in the range of 1mm to 10mm.
- Tolerance of 0.25mm to 1mm should be provided.
- Surface roughness of 1.575mm to 12.3mm can be achieved.

## Economic Considerations

- Slow production rate of 0.5 to 6 pieces per hour.
- High labor skills required.
- Low equipment costs, moderate tooling costs.
- Economical for batch production.

## Quality Considerations

- Explosives contaminate the transfer media, which need to be cleaned each time for better quality and process control.
- It is difficult to control the mechanical properties and dimensions of the part.

### Advantages

- Complex shapes can be produced.
- Smoothness of the contour can be controlled.
- Cheap alternative to Superplastic Forming.

### Disadvantages

- Only viable for low production volumes.
- Handling explosives requires great care and safety precautions.

## Electromagnetic Forming

Electromagnetic forming is a non-contact forming technique. It is an impulse or high-velocity forming method using pulsed magnetic field by application of Lorentz' forces to the work piece preferably made of highly electrically conductive material without having working medium and mechanical contact. Therefore, hollow profiles can be compressed or expanded as well as 2D or 3D sheet metal can be shaped, joined or cut. Electromagnetic forming process has very high strain rate and velocity in comparison to conventional process. Therefore, the Forming limits can be extended for several materials.

Electromagnetic forming (EMF) is also called impulse forming process.

During the EMF process, intense transient magnetic field produced around coil when the discharge current runs through it. According to electromagnetic induction law of Faraday, when the metallic object placed in magnetic field, eddy current will be produced. Two opposite direction of flowing current created a large magnitude magnetic repulsion force between the coil and the metal object (work piece). This force is used to form work piece at very high speed Mamalis, Corriea. Maxwell described physical effects which is the basic of EMF principle. To deform solid conductors Kapitza generate sufficient magnetic field strengths. Thus he provided the foundation for the EMF process. This principle is technologically exploiting for a target-oriented forming of metals. Different applications and detail description is given by Brower. By using electromagnetic forming tubular components or hollow profiles can compressed, expand (i.e. bulge), initially flat or three-dimensional sheet metals can be formed. On the basis of these three different processes variants, types of coils for EMF can be distinguished. In the expansion, tubular work piece around the coil while in compression the coil encloses the work piece. Electromagnetic forming with direct electrode contact is a variant apart from these three major process variants. It is mentioned in Furth and Waniek. Current flow in the work piece is induced by magnet in most cases. While Furth and Waniek claimed that method in which current directly passing through is more efficient than the conventional procedure. They introduced the idea of electromagnetic forming by pulling. While in the most of the applications the work piece is always pushed away from the tool. They suggest a special setup in order to establish pulling force. Another special variant is mentioned in Brower in which the elastic medium is used to apply electromagnetic force. He also established the set up for electromagnetic forming in which he used pressure concentrator and an elastomeric punch, which is placed between the tool coil and work piece. This process is not limited to electrically conductive material. A comparison between electromagnetic forming through an elastic medium and direct electromagnetic forming had given in Livshitz. Only the electromagnetic compression has advantages compared to conventional forming processes at all. A kind of renaissance can be observed over the last years in electromagnetic forming, which is related to implementing lightweight construction concepts in the automotive industry. The electromagnetic sheet metal forming does not have that much significance with in industry but is expected to be increased. The electromagnetic forming process has many advantages in comparison to conventional forming processes. In EMF process the ductility of work piece increase significantly as compare to conventional forming process, though the reason behind this increase in ductility is still being debated.

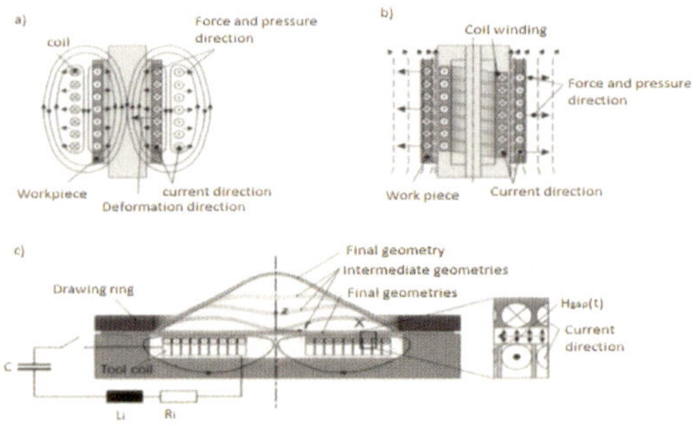


Figure: The magnetic field lines, direction of current and the pressure during a) electromagnetic compression; b) electromagnetic expansion; c) electromagnetic sheet metal forming.

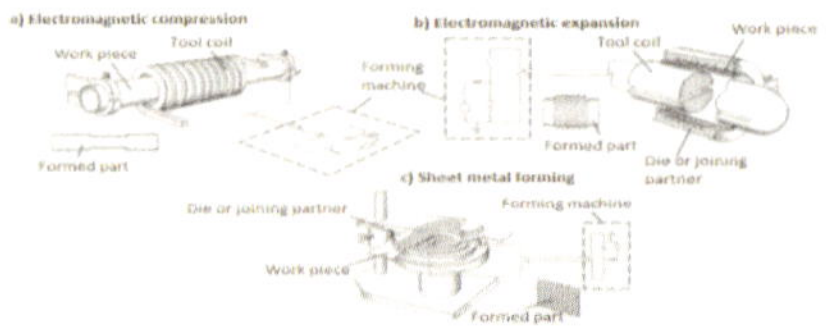

Figure: Different coil types for the EMF processes.

## Principal of Electromagnetic Forming

The equivalent circuit diagram for the setup of electromagnetic forming is shown in fig-3. To form highly electrically conductive material a magnetic field is required, which is achieved by pulse generator. There were so many circuit diagrams of this setup given in literature but fig-3 is the most descriptive design of circuit. In this circuit capacitor C, inductance $L_i$ and resistor $R_i$ are in the series, representing as a forming machine. The tool coil $R_{coil}$ and its inductance $L_{coil}$ both are connected to the pulse power generator in series.

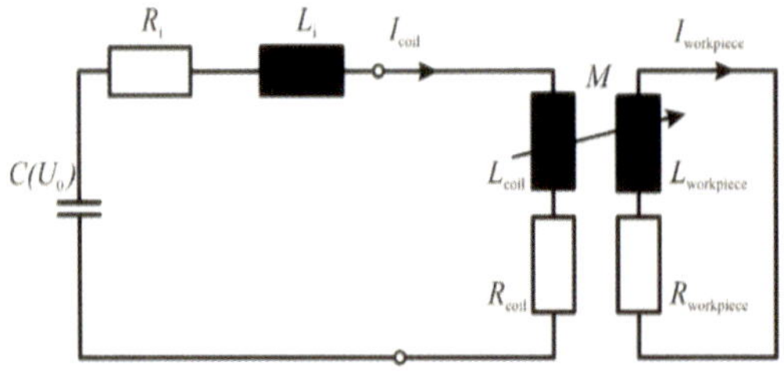


Figure: Detailed version of equivalent circuit diagram

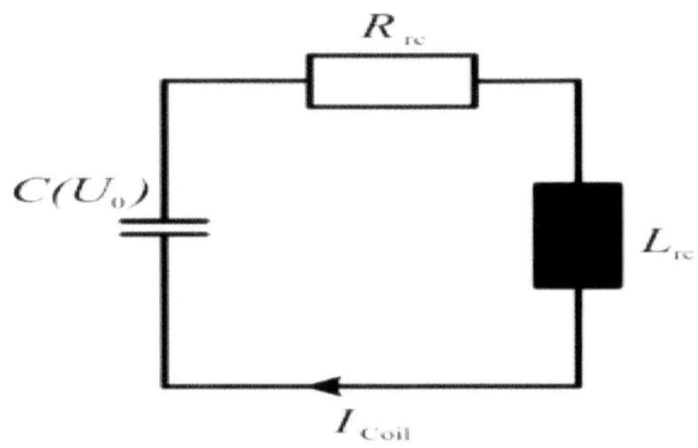


Figure: Reduced version

Capacitor battery stores the energy. The charging energy $E_c$ of this capacitor battery is calculated from the charging voltage $U(t)$ and the capacity C.

$$E_c(t) = \frac{1}{2} CU(t)^2$$

Magnetic pressure acts only in areas of work piece close to coil. Therefore the deformation of work piece starts in these areas with velocity 102 m/s. The different directions of current, magnetic field lines, and magnetic pressure are shown in Figure. The setup is complimented by an additional component in tube compression process. This component is known as field shaper. In some publications it is called field concentrator. Firstly Babat and Losinsky used field shaper to heat work piece inductively. In comparison to direct acting coil, the mechanical loading of tool coil can be significantly reduced by using field shaper, resulting in a higher coil tool life. Combination of inductive heating and electromagnetic forming can be used. To produce a more uniform magnetic field and pressure for sheet metal forming Kamal and Daehn used a field shaper. During electromagnetic tube compression to support the work piece a mandrel can be placed inside the work piece.

Electromagnetic tube compression: Volume force can be determined on the basis of the Magnetic flux density $\vec{B}$ and the current density $\vec{\ }$. The flux density is the result of product of the magnetic field strength H and permeability μ. The negative derivative of magnetic field strength H with respect to the radius r equal to current density $\vec{J}$. The radial forces acting on a tubular work piece $F_r$ can be calculated. These volume forces can be transformed to virtual surface forces mathematically. It is called pressure P. The pressure difference between two points on work piece can be calculated by integrating acting force, using inner radius $r_i$ and outer radius $r_o$ as integration limits.

$$\vec{F} = \vec{J} \times \vec{B}$$

$$\vec{J} = -\frac{\partial H}{\partial r}$$

$$F_r = -\mu H \frac{\partial H}{\partial r} = -\frac{1}{2}\mu \frac{\partial (H^2)}{\partial r}$$

$$p(r,t) = \int \frac{ri}{ro} F(r,t) dr = \frac{1}{2}\mu \left[ H_{gap}{}^2(t) - H_{pen}{}^2(t) \right]$$

In both tube and sheet forming process the magnetic pressure depends only on the penetrated magnetic field $H_{pen}$ and on the strength of magnetic field in the gap between work piece and the tool coil. $H_{gap}$. Using suitable measuring loops these field strengths can be measured in tube compression process. The axial distribution of the field Strength in the gap $H_{gap}$ for a direct acting compression coil setup can be calculated. For this a relationship based on the coil length $l$, effective gape $a$, the coil current $I(t)$ and the number of turns of the tool coil $n$ is required. It is called distribution coefficient $K_H$.

$$H_{gap}(t) = \frac{nl(t)}{l} K_H$$

$$K_H = \frac{1}{\pi}(\arctan\frac{2z+l}{a} - \arctan\frac{2z+l}{a})$$

$$a = a_{Air} + \frac{1}{2}\left(S_{wp} + \delta_{coil}\right)$$

Here $a$ is the effective gap, $a_{Air}$ is the width of the air gap and $\delta_{wp}$ is the skin depth of the work piece and $\delta_{coil}$ is the skin depth of the coil. In order to analyse the effect of field shaper having trapezoidal cross-section on the pressure distribution Yu used two dimensional FE simulations. Finally a dimensional FE simulation was presented by Bahmani. During the first half wave of the coil current, the deformation process completes with the first pressure pulse. Rajive had simulated the electromagnetic compression forming, carried out on steel tubes of 2.3 mm wall thickness and 76.2mm outer diameter. They had compared the experimental values of electromagnetic formed tubes available in literature with results of simulation. Haiping investigated the tendency of homogeneous radial deformation during electromagnetic compression of aluminium tube through sequential coupling numerical simulation and experiments. They found that the tendency depends on the length ratio of tube to coil (R). The discharge length, corresponding to the homogeneous radial deformation is insensitive to discharge voltage. R has critical value $R_c$. The value of $R_c$ can be zero or 1 and it can be increased with the coil length also. They concluded that when- 1) $R > R_c$, the final tube presents horn shape. 2) $R > R_c$, the final tube presents drum shape.

## Electromagnetic Tube Expansion

In the electromagnetic forming process the field strength in the gap depends on the current induced in the work piece. This induced current in tube expansion process is not as high as the coil current. Therefore, the magnetic strength cannot be calculated from the coil current. In order to determine the force acting on the work piece, the magnetic field between the tool coil and work piece is measured. To determine the acting loads on the basis of the measured current in the work piece and measured coil current, Gourdin used principle of virtual work. In expansion process the geometric stiffness decreases while in compression process stiffness increase over the process time. The influence of inertia forces is higher in case of lower stiffness. Thomas introduced a full coupled Lagrangian least action variation principle considering magnetic flux, electric potentials and the displacement field as independent variable. This variation principle provides an efficient staggered solution algorithm, combining with the vibration integration numerical scheme. Zhong had employed finite code ANSYS for the simulation of the electromagnetic tube bulging process. To obtain the radial and axial magnetic pressure acting on the tube ANSYS/EMAG was used. The time space distribution of magnetic pressure was presented on various tube lengths. Effect of tube size on the high velocity deformation, distribution of radial magnetic pressure and axial magnetic pressure were discussed. The axial magnetic pressure of various sizes of tube found to be zero at the centre, greater at the end points (when the tube length is smaller or equal to the coil length), greater at the location equal to the coil length and decreased to the adjacent region (when the tube length is bigger than the coil length). They found that the deformation along the length and radial magnetic pressure distribution changes with dimensionless tube length. Thomas had applied the concept of forming limit diagrams (FLDs) to study the ductility of freely expanding electromagnetically loaded aluminium tubes. Necking strains of various tubes geometries were measured which were loaded by different coils and currents. They plotted experimental results in principle strain space show reasonable agreement with the corresponding theoretical FLD predictions. It was indicated that forming limits increase 2- to 3-fold with respect to the quasi static case.

## Electromagnetic Sheet Metal Forming

The negative derivative of magnetic field strength H with respect to the z- coordinate is equal to current density $\vec{J}$. In electromagnetic sheet metal forming, force acting on a sheet metal work piece can be calculated by eq- 10. The magnetic pressure can be calculated by integrating the acting force over the distance.

$$\vec{J} = -\frac{\partial H}{\partial z}$$

$$Fz = -\mu H \frac{\partial H}{\partial z} = -\frac{1}{2}\frac{\partial (H^2)}{\partial z}$$

During deformation increase of the gap volume between the work piece and tool coil influence the acting magnetic pressure, this effect is much more relevant in case of sheet metal forming than the tube compression or expansion. Therefore the mathematical description of the process in close form is not possible. Equation had given regarding the magnetic pressure $p_r$ and the magnetic field $\vee$ in the radial direction. A harmonic field simulation is used to determine the pressure distribution. The ratio of the field strength $H_{gap}(t_{max'}r)$ (when the current reaches the maximum) to the maximum possible field strength $H_o(t_{max})$ is known as factor of pressure distribution $K_H(r)$, a analogous to the coefficient for the tube compression.

$$K_H(r) = \frac{H_{gap}(t_{max,}r)}{H_o(t_{max}r)} \text{With } H_o = \frac{nl(t_{max}r)}{l_{coil}}$$

During sheet metal deformation process an increase of gap volume decreases the magnetic field Strength therefore a continuously decreasing pressure can be observed. After some deformation of work piece pressure decreases to a negligible level in this situation only inertia force drives the process. Correia (2008) studied the effect of electromagnetic and mechanical forces, produced during sheet metal forming process. Problem was analyzed in ABAQUS/Explicit. They discussed the electromagnetic forming in two ways, first one deals with the formulation and computation of electromagnetic parameters and the second one deals with a FES. They calculated the parameters such as Electromagnetic field density, electromagnetic force distribution and coil discharge by developing the numerical and analytical tool. Unger formulated a thermo-magneto-mechanical multifield model and used to simulate electromagnetic sheet metal forming processes (EMF) based on Lorentz force principle. After simulation carried on a sheet metal made of aluminium alloy AA6005, the deformation results of the sheet metal are obtained. Wei studied Effect of temperature on plastic deformation of sheet by electromagnetic force. They used a new method of controlling welding distortion with trailing electromagnetic force. They studied the plastic deformation of sheet using electromagnetic force at different temperature. It was simulated by ANSYS. They employed Physical environment method to study the electromagnetic force acting on sheet in the presence of high temperature and deformation of sheet. They concluded following results: (1) The magnitude of peak axial force decreases with increasing temperature, on the other hand the magnitude of peak radial force increases with increasing temperature. The variation of peak electromagnetic force with temperature is nonlinear. (2)When temperature increases electromagnetic force decreases due to decrease of yield strength. (3) They proposed that, the electromagnetic force is employed to control welding stain during welding.

Chu had proposed a new method and implemented it to determine the flow curve of aluminium alloy1100-O at high strain rates. By combining retrieved effective stress from simulation and measured effective strain from deformed work piece. The accuracy of determining flow curve is increased by iteration procedure. They had simulated forming process by using determined flow curves. They observed that the deviation in strain could be reduced from 17.9% to 6.74% and effective strain of 0.56 could be achieved by deforming material without failure. Xiao-hui had analyzed the distribution of magnetic forces and current on sheet and coil. They proposed reason for coil failure. To analyze the deformation law of the sheet they had given the magnetic forces on the sheet as input into software ANSYS/LS-DYNA. They concluded that by increasing distance between the coil and the sheet, the distribution of magnetic forces on the uniform pressure electromagnetic actuator is changed which may cause coil failure. Okoye applied a hybrid forming process for the formation of metals based on Lorentz force termed as EMAS (electromagnetic-assisted stamping). This technology was applied under consideration of making fuel efficient cars and aircrafts, large size vehicles panels, and improved formability limit of materials. They gave more attention on designing of coil such as pressure distribution across the coil. The coils were finally configured to bend in order to form shapes, single, double or more curved. The design was optimized subjected to maximum electrical efficiency to increase the durability of the coil.

## Energy Interaction during Electromagnetic Forming

In electromagnetic forming process initially the energy stored in the pulsed power generator as capacitive energy. This energy is used to deform the work piece at the end.

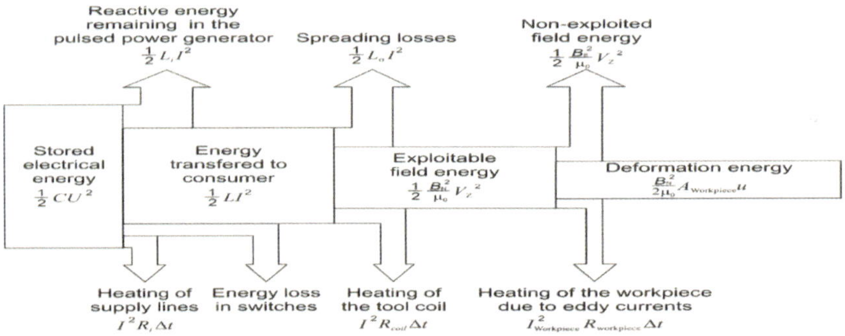


Figure: Energy transfer during electromagnetic forming.

A complex energy transfer is there in between these two stages. Only parts of stored energy transfer to the coil and used to deform work piece. There are losses in between energy storage and deformation of work piece. They mentioned the range of efficiency of electromagnetic forming 10% to 40%. Psyk was focused on the effect of energy transfer on the forming results. The Capacitor charging energy $E_c$ at first is transferred into a magnetic pressure pulse p. This magnetic pressure pulse is transferred into kinetic energy $E_{kin}$ and then transferred into forming energy $E_{def}$. During deformation process there is a moment when pressure has decreased to zero, the kinetic energy stored in the work piece is used to finish the forming process. He performed two exemplary forming tasks. In first task a die with a spherical cavity used to form an aluminium sheet and in the second task a conical die is used to form aluminium sheet. The total efficiency of process found to be 23% in first task and 28% in second task.

## Electroforming

Electroforming is a metal forming process that uses electrodeposition to form, or grow, metal parts onto a model, which is known as a "mandrel." Electrodeposition involves the electrochemical transfer of metal ions through an electrolyte to a surface from an anode.

The mandrel determines the shape of the final product, which may be referred to as an "electroform." This mandrel is conditioned such that the metal being deposited does not adhere to it. After the process is complete, the mandrel is typically removed. The outer surface of the mandrel creates the inner surface of the finished product and the metal that formed around the mandrel retains the shape created by electrode position. This metal form is the finished product.

This technique allows for high accuracy and the creation of complex shapes. The surface of the form that was touching the mandrel is highly accurate with respect to the desired shape of the model. The outside areas won't be as well-defined, but you can add as much thickness as needed to the piece on the outside layers. The metal layers can be extremely thin if desired. It only needs to be thick enough to support itself.

### Working of Electroforming Process

The electroforming process begins with a metal substrate with a textured surface — the mandrel. You must clean and degrease the substrate and then apply a photo-sensitive coating. You then cover the areas to which you do not want the metal ions to adhere with a film photomask. This is transferred to the surface through exposure to ultraviolet light, creating a "negative" of the final product.

Next comes the electrodeposition phase of the process, which is what causes the metal to form on the mandrel.

Electrodeposition is, at its most basic, the method used to deposit metal onto a conductive surface. This process involves immersing an anode and a cathode in an electrolytic bath made of a solution of salt and the metal that will be deposited onto the object. You pass a direct electric current (DC) through the salt-and-metal solution, which transforms the metallic ions into atoms and causes them to deposit onto the cathode, which is the object that will receive the metal. This allows you to build up the object essentially atom by atom, allowing for the high levels of accuracy for which electroforming is known.

In electroforming, the atoms will only adhere to the parts not masked with the film photomask. One the metal has grown on the mandrel to the desired thickness, you remove it from the electrolytic bath. If you're going to remove the mandrel, as is the case with a majority of projects, you will either keep it intact and separate it or dissolve it away. Sometimes — especially if the electroforming serves a decorative purpose — you will leave the mandrel in place.

There are two different methods of electroforming used to control the shape of the finished product: Electroforming overgrowth and electroforming thick-resist.

Electroforming overgrowth is plating-defined and uses a thin photoresist pattern that covers parts of the conductive substrate that will polymerize under UV light. The metal forms over the photoresist pattern and is allowed to grow over the sides of the model. The size of the apertures depends on the overgrowth of the electroforming process, which means the plating process defines the product's accuracy. This method is commonly used to create items such as nozzles and filters due to the bellmouth shapes it can form.

The thick-resist method is photo-defined because its accuracy is defined by the photolithographic process. It involves depositing the metal material below the height of the photoresist. No overgrowth is allowed, resulting in straight-lined walls. This type of model creates a two-dimensional structure with straight sidewalls and you can also apply multiple cycles of the process to create quasi-three-dimensional structures. The height of the photoresist determines the height of the electroform.

## Electroforming with Non-conductive Materials

Up to this point, we've covered electroforming with conductive metals as a substrate. This required blocking off the parts of the mandrel where you did not want metal to form. You can also use electroforming with non-conductive materials such as plastics and glass.

The process of electroforming with a non-conductive mandrel is similar to the process involving a conductive mandrel, but you have to do the beginning steps in an opposite manner. To clarify, instead of covering the parts of the surface where you don't want metal to form, you have to make the areas where you do want coverage conductive so that metal will grow there. You do this by applying a conductive coating to the desired parts of the model, which you can do manually, chemically or with vacuum deposition techniques. Manual methods are common and include painting or spraying on a conductive metallic paint that creates a thin film on the substrate, often comprised of nickel or silver.

Once the conductive film is applied to the mandrel, you can go through the electroforming process as you would with a conductive substrate.

## Electroplating vs. Electroforming

Electroforming is often confused with a process called "electroplating." The processes have some similarities but produce different results and have different applications.

Electroforming and electroplating are both used in the manufacturing of metal products and they both rely on the electrodeposition process. While electroforming creates a product that is separate from the model, the metal deposited in electroplating becomes part of the surface it's deposited on. A product made by electroforming is usually created entirely through electroforming, while electroplating forms just the outer layer of a product.

Electroplating is used to deposit a thin coating on the outside surface of an object, typically to enhance its appearance or functionality in some way. You might use electroplating to improve the cosmetic appeal of an object or its hardness, resistance to corrosion, conductivity or many other things. The process can be applied to a wide range of objects, including electroforms, and the plating can be made of a variety of metals.

In short, the main similarities between electroforming and electroplating are as follows:

- They are both used in the manufacturing of metal objects.
- They both use the electrodeposition process.
- You can use a variety of metals for both processes.

The main difference between the two is that electroforming creates a separate object, while electroplating deposits a layer onto an existing object.

## Advantages of Electroforming

There are several advantages when you choose electroforming for your product:

- High Accuracy: You can electroform metals to extremely tight tolerances of up to one micrometer. The fact that you create the metal layer atom by atom allows for extremely fine details on its surface and extremely small components. You also won't see any tool marks or shrinkage like you would with other metal forming techniques.
- Complexity: Electroforming lends itself well to producing complex shapes because the electroform is created from the negative image of the mandrel. This makes electroforming useful for a wide variety of applications. Many of the shapes that are difficult to create via other methods are easily produced with electroforming. The method is especially useful for accurately manufacturing internal cavities.
- Light Weight: The walls of an electroformed object can be extremely thin. They only need to be thin enough that they hold together and won't be damaged by regular handling. These thin components are also lightweight, which is especially useful in the automotive and aerospace fields.
- Metal Purity: The way the metal layer forms, atom by atom, results in extremely pure metal. This purity increases the quality of the metal and characteristics such as hardness, malleability and conductivity.
- Highly Flexible: Electroforming is a relatively flexible metal manufacturing technique and has a wide range of applications across industries. You can make extremely small components as well as large ones. The size of the item is only limited by the size of the plating tank.

It's also useful for both low- and high-volume production runs. You can use a permanent mandrel for high-volume production runs. For short runs, you can use a disposable mandrel that you melt or dissolve away from the electroform.

You can also incorporate other components into the product of the initial electroforming process and add parts onto after this first step as necessary. It's also possible to join two components via electroforming. After the electroforming process is completed, other metal finishing processes, such as plating, can increase the quality of the item and add or improve properties such as hardness, corrosion-resistance and conductivity.

### Disadvantages and Challenges

The outside surface of the electroform is often less precise than the inside part of the form, especially when the overgrowth method is used. The thicker the object, the more definition is lost. Excess material can sometimes build up on outside corners and edges, but this unwanted material can be etched away.

Certain challenges can also sometimes arise during the electroforming process — especially when removing the mandrel from the finished part. The final product might also not have all the properties you need, but this problem can often be resolved by plating the object, especially with copper or palladium.

## Hydroforming

Hydroforming is a metal fabricating and forming process which allows the shaping of metals such as steel, stainless steel, copper, aluminum, and brass. This process is a cost-effective and specialized type of die molding that utilizes highly pressurized fluid to form metal. Generally there are two classifications used to describe hydroforming, sheet hydroforming and tube hydroforming. Sheet hydroforming uses one die and a sheet of metal; the blank sheet is driven into the die by high pressure water on one side of the sheet forming the desired shape. Tube hydroforming is the expansion of metal tubes into a shape using two die halves, which contain the raw tube. Hydroforming is used to replace the older process of stamping two part halves and welding them together. It is also used to make parts both more efficiently by eliminating welding as well as creating complex shapes and contours. Parts created in this method have a number of manufacturing benefits including seamless bonding, increased part strength, and the ability to maintain high-quality surfaces for finishing purposes. When compared to traditional metal stamped and welded parts, hydroformed parts are lightweight, have a lower cost per unit, and are made with a higher stiffness to weight ratio. The processes can also be utilized in the single stage production of components; saving labor, tools, and materials.

### References


- Sheet-metal-forming: custompartnet.com, Retrieved 29 July 2018
- Forging-process: themetalcasting.com, Retrieved 06 May 2018
- Extrusion-process: themetalcasting.com, Retrieved 16 March 2018
- Explosive-forming: engineeringclicks.com, Retrieved 25 April 2018
- Electroforming-guide: sharrettsplating.com, Retrieved 26 June 2018
- What-is-hydroforming: americanhydroformers.com, Retrieved 19 April 2018

6

# Welding and its Types

Welding is a sculptural or fabrication process that is used to join materials like thermoplastics or metals through the process of fusion. This chapter discusses in detail the different methods and processes of welding such as arc welding, oxy-fuel welding and cutting, shielded metal arc welding, gas tungsten arc welding, gas metal arc welding, flux-cored arc welding, etc. which are commonly used in manufacturing.

Welding is a technique which is used for joining metallic parts usually through the application of heat.

This technique was discovered during efforts to manipulate iron into useful shapes. Welded blades were developed in the 1st millennium CE, the most famous being those produced by Arab armourers at Damascus, Syria. The process of carburization of iron to produce hard steel was known at this time, but the resultant steel was very brittle. The welding technique—which involved interlayering relatively soft and tough iron with high-carbon material, followed by hammer forging—produced a strong, tough blade.

In modern times, the improvement in iron-making techniques, especially the introduction of cast iron, restricted welding to the blacksmith and the jeweler. Other joining techniques, such as fastening by bolts or rivets, were widely applied to new products, from bridges and railway engines to kitchen utensils.

Modern fusion welding processes are an outgrowth of the need to obtain a continuous joint on large steel plates. Rivetting had been shown to have disadvantages, especially for an enclosed container such as a boiler. Gas welding, arc welding, and resistance welding all appeared at the end of the 19th century. The first real attempt to adopt welding processes on a wide scale was made during World War I. By 1916 the oxyacetylene process was well developed, and the welding techniques employed then are still used. The main improvements since then have been in equipment and safety. Arc welding, using a consumable electrode, was also introduced in this period, but the bare wires initially used produced brittle welds. A solution was found by wrapping the bare wire with asbestos and an entwined aluminum wire. The modern electrode, introduced in 1907, consists of a bare wire with a complex coating of minerals and metals. Arc welding was not universally used until World War II, when the urgent need for rapid means of construction for shipping, power plants, transportation, and structures spurred the necessary development work.

Resistance welding, invented in 1877 by Elihu Thomson, was accepted long before arc welding for spot and seam joining of sheet. Butt welding for chain making and joining bars and rods was developed during the 1920s. In the 1940s the tungsten-inert gas process, using a nonconsumable tungsten electrode to perform fusion welds, was introduced. In 1948 a new gas-shielded process utilized a wire electrode that was consumed in the weld. More recently, electron-beam welding, laser welding, and several solid-phase processes such as diffusion bonding, friction welding, and ultrasonic joining have been developed.

## Basic Principles of Welding

A weld can be defined as a coalescence of metals produced by heating to a suitable temperature with or without the application of pressure, and with or without the use of a filler material.

In fusion welding, a heat source generates sufficient heat to create and maintain a molten pool of metal of the required size. The heat may be supplied by electricity or by a gas flame. Electric resistance welding can be considered fusion welding because some molten metal is formed.

Solid-phase processes produce welds without melting the base material and without the addition of a filler metal. Pressure is always employed, and generally some heat is provided. Frictional heat is developed in ultrasonic and friction joining, and furnace heating is usually employed in diffusion bonding.

The electric arc used in welding is a high-current, low-voltage discharge generally in the range 10–2,000 amperes at 10–50 volts. An arc column is complex but, broadly speaking, consists of a cathode that emits electrons, a gas plasma for current conduction, and an anode region that becomes comparatively hotter than the cathode due to electron bombardment. A direct current arc is usually used, but alternating current arcs can be employed.

Total energy input in all welding processes exceeds that which is required to produce a joint, because not all the heat generated can be effectively utilized. Efficiencies vary from 60 to 90 percent, depending on the process; some special processes deviate widely from this figure. Heat is lost by conduction through the base metal and by radiation to the surroundings.

Most metals, when heated, react with the atmosphere or other nearby metals. These reactions can be extremely detrimental to the properties of a welded joint. Most metals, for example, rapidly oxidize when molten. A layer of oxide can prevent proper bonding of the metal. Molten-metal droplets coated with oxide become entrapped in the weld and make the joint brittle. Some valuable materials added for specific properties react so quickly on exposure to the air that the metal deposited does not have the same composition as it had initially. These problems have led to the use of fluxes and inert atmospheres.

In fusion welding the flux has a protective role in facilitating a controlled reaction of the metal and then preventing oxidation by forming a blanket over the molten material. Fluxes can be active and help in the process or inactive and simply protect the surfaces during joining.

Inert atmospheres play a protective role similar to that of fluxes. In gas-shielded metal-arc and gas-shielded tungsten-arc welding an inert gas—usually argon—flows from an annulus surrounding the torch in a continuous stream, displacing the air from around the arc. The gas does not chemically react with the metal but simply protects it from contact with the oxygen in the air.

The metallurgy of metal joining is important to the functional capabilities of the joint. The arc weld illustrates all the basic features of a joint. Three zones result from the passage of a welding arc: (1) the weld metal, or fusion zone, (2) the heat-affected zone, and (3) the unaffected zone. The weld metal is that portion of the joint that has been melted during welding. The heat-affected zone is a region adjacent to the weld metal that has not been welded but has undergone a change in microstructure or mechanical properties due to the heat of welding. The unaffected material is that which was not heated sufficiently to alter its properties.

Weld-metal composition and the conditions under which it freezes (solidifies) significantly affect the ability of the joint to meet service requirements. In arc welding, the weld metal comprises filler material plus the base metal that has melted. After the arc passes, rapid cooling of the weld metal occurs. A one-pass weld has a cast structure with columnar grains extending from the edge of the molten pool to the centre of the weld. In a multipass weld, this cast structure may be modified, depending on the particular metal that is being welded.

The base metal adjacent to the weld, or the heat-affected zone, is subjected to a range of temperature cycles, and its change in structure is directly related to the peak temperature at any given point, the time of exposure, and the cooling rates. The types of base metal are too numerous to discuss here, but they can be grouped in three classes: (1) materials unaffected by welding heat, (2) materials hardened by structural change, (3) materials hardened by precipitation processes.

Welding produces stresses in materials. These forces are induced by contraction of the weld metal and by expansion and then contraction of the heat-affected zone. The unheated metal imposes a restraint on the above, and as contraction predominates, the weld metal cannot contract freely, and a stress is built up in the joint. This is generally known as residual stress, and for some critical applications must be removed by heat treatment of the whole fabrication. Residual stress is unavoidable in all welded structures, and if it is not controlled bowing or distortion of the weldment will take place. Control is exercised by welding technique, jigs and fixtures, fabrication procedures, and final heat treatment.

There are a wide variety of welding processes. Several of the most important are discussed below.

## Forge Welding

This original fusion technique dates from the earliest uses of iron. The process was first employed to make small pieces of iron into larger useful pieces by joining them. The parts to be joined were first shaped, then heated to welding temperature in a forge and finally hammered or pressed together. The Damascus sword, for example, consisted of wrought-iron bars hammered until thin, doubled back on themselves, and then rehammered to produce a forged weld. The larger the number of times this process was repeated, the tougher the sword that was obtained. In the Middle Ages cannons were made by welding together several iron bands, and bolts tipped with steel fired from crossbows were fabricated by forge welding. Forge welding has mainly survived as a blacksmith's craft and is still used to some extent in chain making.

## Arc Welding

Shielded metal-arc welding accounts for the largest total volume of welding today. In this process an electric arc is struck between the metallic electrode and the workpiece. Tiny globules of molten metal are transferred from the metal electrode to the weld joint. Since arc welding can be done with either alternating or direct current, some welding units accommodate both for wider application. A holder or clamping device with an insulated handle is used to conduct the welding current to the electrode. A return circuit to the power source is made by means of a clamp to the workpiece.

Gas-shielded arc welding, in which the arc is shielded from the air by an inert gas such as argon or helium, has become increasingly important because it can deposit more material at a higher efficiency and can be readily automated. The tungsten electrode version finds its major applications in highly alloyed sheet materials. Either direct or alternating current is used, and filler metal is added either hot or cold into the arc. Consumable electrode gas-metal arc welding with a carbon dioxide shielding gas is widely used for steel welding. Two processes known as spray arc and short-circuiting arc are utilized. Metal transfer is rapid, and the gas protection ensures a tough weld deposit.

Submerged arc welding is similar to the above except that the gas shield is replaced with a granulated mineral material as a flux, which is mounded around the electrode so that no arc is visible.

Plasma welding is an arc process in which a hot plasma is the source of heat. It has some similarity to gas-shielded tungsten-arc welding, the main advantages being greater energy concentration, improved arc stability, and easier operator control. Better arc stability means less sensitivity to joint alignment and arc length variation. In most plasma welding equipment, a secondary arc must first be struck to create an ionized gas stream and permit the main arc to be started. This secondary arc may utilize either a high-frequency or a direct contact start. Water cooling is used because of the high energies forced through a small orifice. The process is amenable to mechanization, and rapid production rates are possible.

## Thermochemical Processes

One such process is gas welding. It once ranked as equal in importance to the metal-arc welding processes but is now confined to a specialized area of sheet fabrication and is probably used as much by artists as in industry. Gas welding is a fusion process with heat supplied by burning acetylene in oxygen to provide an intense, closely controlled flame. Metal is added to the joint in the form of a cold filler wire. A neutral or reducing flame is generally desirable to prevent base-metal oxidation. By deft craftsmanship very good welds can be produced, but welding speeds are very low. Fluxes aid in preventing oxide contamination of the joint.

Another thermochemical process is aluminothermic (thermite) joining. It has been successfully used for both ferrous and nonferrous metals but is more frequently used for the former. A mixture of finely divided aluminum and iron oxide is ignited to produce a superheated liquid metal at about 2,800°C (5,000°F). The reaction is completed in 30 seconds to 2 minutes regardless of the size of the charge. The process is suited to joining sections with large, compact cross sections, such as rectangles and rounds. A mold is used to contain the liquid metal.

## Resistance Welding

Spot, seam, and projection welding are resistance welding processes in which the required heat for joining is generated at the interface by the electrical resistance of the joint. Welds are made in a relatively short time (typically 0.2 seconds) using a low-voltage, high-current power source with force applied to the joint through two electrodes, one on each side. Spot welds are made at regular intervals on sheet metal that has an overlap. Joint strength depends on the number and size of the welds. Seam welding is a continuous process wherein the electric current is successively pulsed into the joint to form a series of overlapping spots or a continuous seam. This process is used to weld containers or structures where spot welding is insufficient. A projection weld is formed when one of the parts to be welded in the resistance machine has been dimpled or pressed to form a protuberance that is melted down during the weld cycle. The process allows a number of predetermined spots to be welded at one time. All of these processes are capable of very high rates of production with continuous quality control. The most modern equipment in resistance welding includes complete feedback control systems to self-correct any weld that does not meet the desired specifications.

Flash welding is a resistance welding process where parts to be joined are clamped, the ends brought together slowly and then drawn apart to cause an arc or flash. Flashing or arcing is continued until the entire area of the joint is heated; the parts are then forced together and pressure maintained until the joint is formed and cooled.

Low- and high-frequency resistance welding is used for the manufacture of tubing. The longitudinal joint in a tube is formed from metal squeezed into shape with edges abutted. Welding heat is governed by the current passing through the work and the speed at which the tube goes through the rolls. Welding speeds of 60 metres (200 feet) per minute are possible in this process.

## Electron-beam Welding

In electron-beam welding, the workpiece is bombarded with a dense stream of high-velocity electrons. The energy of these electrons is converted to heat upon impact. A beam-focusing device is included, and the workpiece is usually placed in an evacuated chamber to allow uninterrupted electron travel. Heating is so intense that the beam almost instantaneously vaporizes a hole through the joint. Extremely narrow deep-penetration welds can be produced using very high voltages—up to 150 kilovolts. Workpieces are positioned accurately by an automatic traverse device; for example, a weld in material 13 mm (0.5 inch) thick would only be 1 mm (0.04 inch) wide. Typical welding speeds are 125 to 250 cm (50 to 100 inches) per minute.

## Cold Welding

Cold welding, the joining of materials without the use of heat, can be accomplished simply by pressing them together. Surfaces have to be well prepared, and pressure sufficient to produce 35 to 90 percent deformation at the joint is necessary, depending on the material. Lapped joints in sheets and cold-butt welding of wires constitute the major applications of this technique. Pressure can be applied by punch presses, rolling stands, or pneumatic tooling. Pressures of 1,400,000 to 2,800,000 kilopascals (200,000 to 400,000 pounds per square inch) are needed to produce a joint in aluminum; almost all other metals need higher pressures.

## Friction Welding

In friction welding, two workpieces are brought together under load with one part rapidly revolving. Frictional heat is developed at the interface until the material becomes plastic, at which time the rotation is stopped and the load is increased to consolidate the joint. A strong joint results with the plastic deformation, and in this sense the process may be considered a variation of pressure welding. The process is self-regulating, for, as the temperature at the joint rises, the friction coefficient is reduced and overheating cannot occur. The machines are almost like lathes in appearance. Speed, force, and time are the main variables. The process has been automated for the production of axle casings in the automotive industry.

## Laser Welding

Laser welding is accomplished when the light energy emitted from a laser source is focused upon a workpiece to fuse materials together. The limited availability of lasers of sufficient power for most welding purposes has so far restricted its use in this area. Another difficulty is that the speed and the thickness that can be welded are controlled not so much by power but by the thermal conductivity of the metals and by the avoidance of metal vaporization at the surface. Particular applications of the process with very thin materials up to 0.5 mm (0.02 inch) have, however, been very successful. The process is useful in the joining of miniaturized electrical circuitry.

## Diffusion Bonding

This type of bonding relies on the effect of applied pressure at an elevated temperature for an appreciable period of time. Generally, the pressure applied must be less than that necessary to cause 5 percent deformation so that the process can be applied to finished machine parts. The process has been used most extensively in the aerospace industries for joining materials and shapes that otherwise could not be made—for example, multiple-finned channels and honeycomb construction. Steel can be diffusion bonded at above 1,000°C (1,800°F) in a few minutes.

## Ultrasonic Welding

Ultrasonic joining is achieved by clamping the two pieces to be welded between an anvil and a vibrating probe or sonotrode. The vibration raises the temperature at the interface and produces the weld. The main variables are the clamping force, power input, and welding time. A weld can be made in 0.005 second on thin wires and up to 1 second with material 1.3 mm (0.05 inch) thick. Spot welds and continuous seam welds are made with good reliability. Applications include extensive use on lead bonding to integrated circuitry, transistor canning, and aluminum can bodies.

## Explosive Welding

Explosive welding takes place when two plates are impacted together under an explosive force at high velocity. The lower plate is laid on a firm surface, such as a heavier steel plate. The upper plate is placed carefully at an angle of approximately 5° to the lower plate with a sheet of explosive material on top. The charge is detonated from the hinge of the two plates, and a weld takes place in microseconds by very rapid plastic deformation of the material at the interface. A completed weld has the appearance of waves at the joint caused by a jetting action of metal between the plates.

## Weldability of Metals

Carbon and low-alloy steels are by far the most widely used materials in welded construction. Carbon content largely determines the weldability of plain carbon steels; at above 0.3 percent carbon some precautions have to be taken to ensure a sound joint. Low-alloy steels are generally regarded as those having a total alloying content of less than 6 percent. There are many grades of steel available, and their relative weldability varies.

Aluminum and its alloys are also generally weldable. A very tenacious oxide film on aluminum tends to prevent good metal flow, however, and suitable fluxes are used for gas welding. Fusion welding is more effective with alternating current when using the gas-tungsten arc process to enable the oxide to be removed by the arc action.

Copper and its alloys are weldable, but the high thermal conductivity of copper makes welding difficult. Refractory metals such as zirconium, niobium, molybdenum, tantalum, and tungsten are usually welded by the gas-tungsten arc process. Nickel is the most compatible material for joining, is weldable to itself, and is extensively used in dissimilar metal welding of steels, stainlesses, and copper alloys.

## Arc Welding

Arc welding is one of several fusion processes for joining metals. By applying intense heat, metal at the joint between two parts is melted and caused to intermix - directly, or more commonly, with an intermediate molten filler metal. Upon cooling and solidification, a metallurgical bond is created. Since the joining is an intermixture of metals, the final weldment potentially has the same strength properties as the metal of the parts. This is in sharp contrast to non-fusion processes of joining (i.e. soldering, brazing etc.) in which the mechanical and physical properties of the base materials cannot be duplicated at the joint.

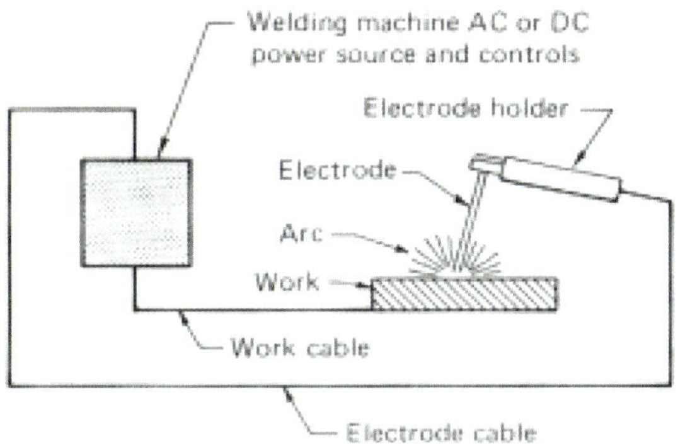


Figure: The basic arc-welding circuit

In arc welding, the intense heat needed to melt metal is produced by an electric arc. The arc is formed between the actual work and an electrode (stick or wire) that is manually or mechanically guided along the joint. The electrode can either be a rod with the purpose of simply carrying the current between the tip and the work. Or, it may be a specially prepared rod or wire that not only conducts the current but also melts and supplies filler metal to the joint. Most welding in the manufacture of steel products uses the second type of electrode.

### Basic Welding Circuit

The basic arc-welding circuit is illustrated in Figure. An AC or DC power source, fitted with whatever controls may be needed, is connected by a work cable to the workpiece and by a "hot" cable to an electrode holder of some type, which makes an electrical contact with the welding electrode.

An arc is created across the gap when the energized circuit and the electrode tip touches the workpiece and is withdrawn, yet still with in close contact.

The arc produces a temperature of about 6500°F at the tip. This heat melts both the base metal and the electrode, producing a pool of molten metal sometimes called a "crater." The crater solidifies behind the electrode as it is moved along the joint. The result is a fusion bond.

## Arc Shielding

However, joining metals requires more than moving an electrode along a joint. Metals at high temperatures tend to react chemically with elements in the air - oxygen and nitrogen. When metal in the molten pool comes into contact with air, oxides and nitrides form which destroy the strength and toughness of the weld joint. Therefore, many arc-welding processes provide some means of covering the arc and the molten pool with a protective shield of gas, vapor, or slag. This is called arc shielding. This shielding prevents or minimizes contact of the molten metal with air. Shielding also may improve the weld. An example is a granular flux, which actually adds deoxidizers to the weld.

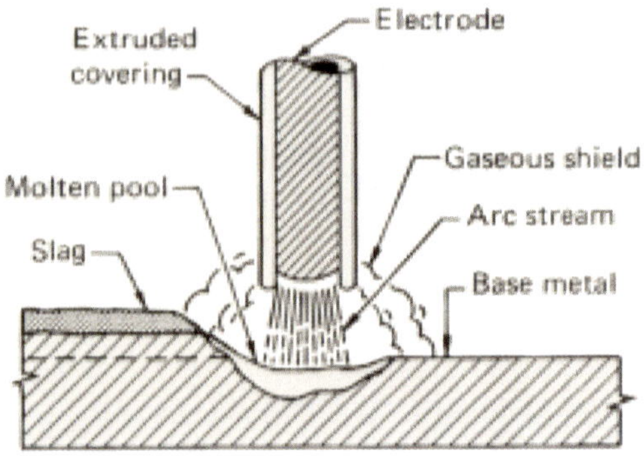


Figure: This shows how the coating on a coated (stick) electrode provides a gaseous shield around the arc and a slag covering on the hot weld deposit.

Figure illustrates the shielding of the welding arc and molten pool with a Stick electrode. The extruded covering on the filler metal rod, provides a shielding gas at the point of contact while the slag protects the fresh weld from the air.

The arc itself is a very complex phenomenon. In-depth understanding of the physics of the arc is of little value to the welder, but some knowledge of its general characteristics can be useful.

## Different Types of Arc Welding

There are different types of arc welding. Which arc welding method you use depends mostly on the metal. Following is an overview of various kinds of arc welding techniques:

## Flux-cored Arc Welding

This type of arc welding uses tubular electrodes filled with flux. While emissive flux shields the arc from air, nonemissive fluxes may need shielding gases. It is ideal for welding dense sections that are an inch or more thick because FCAW has a higher weld-metal deposition rate

## Gas Metal Arc Welding

GMAW or MIG welding shields the arc with a gas like argon or helium or a gas mix. The electrodes have deoxidizers that prevent oxidation, so you can weld multiple layers. This method has several benefits: simple, versatile, economical, low temperatures, and easily automated. This is a popular welding technique for thin sheets and sections.

## Gas Tungsten Arc Welding

GTAW or TIG welding, is often considered to be the most difficult. Tungsten electrodes create the arc. Inert gases like argon or helium or a mix of the two is used to protect the shield. Filler wires add molten material if needed. This method is much "cleaner" as it doesn't produce slag, making it ideal for welding jobs where appearance matters as well as thin materials.

## Plasma Arc Welding

This arc welding technique uses ionized gases and electrodes that create hot plasma jets aimed at the welding area. As the jets are extremely hot, this method for narrow and deep welds. PAW is also good for increasing welding speeds.

## Shielded Metal Arc Welding

SMAW is one of the simplest, oldest, and most adaptable arc welding methods, making it very popular. The arc is generated when the coated electrode tip touches the welding area and is then withdrawn to maintain the arc. The heat melts the tip, coating, and metal, so that the weld is formed once that alloy solidifies. This technique is typically used in pipeline work, shipbuilding, and construction.

## Submerged Arc Welding

SAW works with a granular flux that creates a thick layer during welding, which completely covers the molten metal and prevents sparks and spatter. This method enables deeper heat penetration because it acts like a thermal insulator. SAW is sued for high-speed sheet or plate steel welding. It can be semiautomatic or automatic. However, it is limited to horizontal welds.

## Arc Welding Processes

In arc welding process, an electric arc between an electrode and a workpiece or between two electrodes is utilized to weld base metals. The basic principle of arc welding is shown in below Figure. However, the basic elements involved in arc welding process are shown in the below Figure. Most of these processes use some shielding gas while others use coatings or fluxes to prevent the weld pool from the surrounding atmosphere.

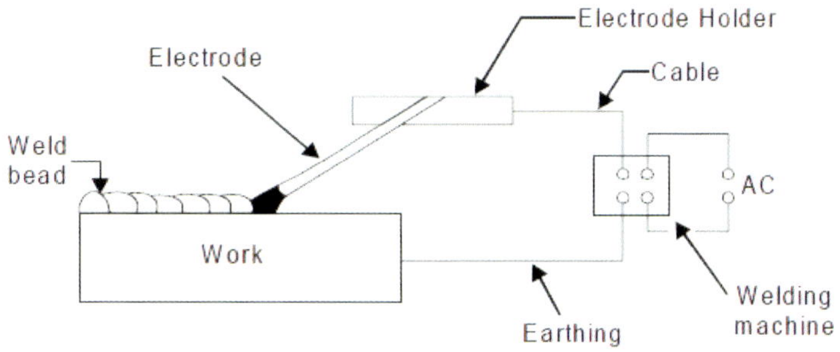


Figure: Principle of arc welding

## Arc Welding Equipment

Arc welding equipment, setup, related tools, and accessories are shown in figure.

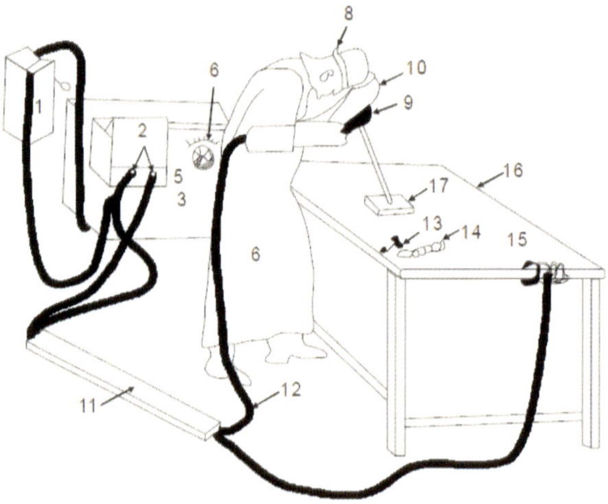


Figure: Arc welding process setup
(1) Switch box.(2) Secondary terminals. (3) Welding machine. (4) Current reading scale. (5) Current regulating hand wheel. (6) Leather apron. (7) Asbestos hand gloves. (8) Protective glasses strap. (9) Electrode holder. (10) Hand shield. (11) Channel for cable protection. (12) Welding cable. (13) Chipping hammer. (14) Wire brush. (15) Earth clamp. (16) Welding table (metallic). (17) Job.

However, some common tools of arc welding are shown below in Figure. Few of the important components of arc welding setup are described as under.

## Arc Welding Power Source

Both direct current and alternating current are used for electric arc welding process, each having its particular applications. DC welding supply is usually obtained from generators driven by electric motor or if no electricity is available by internal combustion engines. For AC welding supply, transformers are predominantly used for almost all arc welding where mains electricity supply is available. They have to step down the usual supply voltage (200-400 volts) to the normally open circuit welding voltage (50-90 volts). The following factors influence the selection of a power source:

1. Type of electrodes to be used and metals to be welded
2. Available power source (AC or DC)
3. Required output
4. Duty cycle
5. Initial costs and running costs
6. Efficiency
7. Available floor space area and
8. Versatility of equipment

## Welding Cables

Welding cables are required for conduction of current from the power source through the electrode holder, the arc, the workpiece and back to the welding power source. These are insulated copper or aluminium cables.

## Electrode Holder

Electrode holder is used for holding the electrode manually and conducting current to it. These are usually matched to the size of the lead, which in turn matched to the amperage output of the arc welder. Electrode holders are available in sizes that range from 150 to 500 Amps.

Electrode Holder

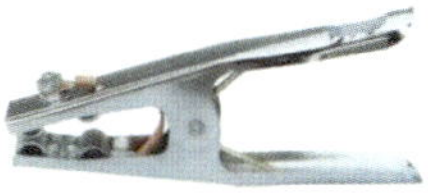

Earth Clamp

V-block

C-clamp

Chippping Hammer

Scriber

Wire Brush

## Welding Electrodes

An electrode is a piece of wire or a rod of a metal or alloy, with or without coatings. An arc is set up between electrode and workpiece. Welding electrodes are classified into the following types:

1. Consumable Electrodes
   i. Bare Electrodes
   ii. Coated Electrodes
2. Non-consumable Electrodes
   i. Carbon or Graphite Electrodes
   ii. Tungsten Electrodes

Consumable electrodes are made up of different metals and their alloys. The end of this electrode starts melting when the arc is struck between the electrode and workpiece. Thus consumable electrode itself acts as a filler metal. Bare electrodes consist of a metal or alloy wire without any flux coating on them. Coated electrodes have flux coating which starts melting as soon as an electric arc is struck. This coating on melting performs many functions like prevention of joint from atmospheric contamination, arc stabilizers etc.

Non-consumable electrodes are made up of high melting point materials like carbon, pure tungsten or alloy tungsten etc. These electrodes do not melt away during welding. But practically, the electrode length goes on decreasing with the passage of time, because of oxidation and vaporization of the electrode material during welding. The materials of non-consumable electrodes are usually copper coated carbon or graphite, pure tungsten, thoriated or zirconiated tungsten.

### Chipping Hammer

Chipping Hammer is used to remove the slag by striking.

- Wire brush

  Wire brush is used to clean the surface to be weld.

- Hand Screen

  Hand screen is used for protection of eyes and supervision of weld bead.

- Protective clothing

  The operator wears the protective clothing such as apron to protect the body from the exposure of direct heat to the body.

## Oxy-fuel Welding and Cutting

Oxy-fuel welding, commonly referred to as oxy welding or gas welding is a process of joining metals by application of heat created by gas flame. The fuel gas commonly acetylene, when mixed with proper proportion of oxygen in a mixing chamber of welding torch, produces a very hot flame of about 5700-5800°F. With this flame it is possible to bring any of the so-called commercial metals, namely: cast iron, steel, copper, and aluminum, to a molten state and cause a fusion of two pieces of like metals in such a manner that the point of fusion will very closely approach the strength of the metal fused. If more metal of like nature is added, the union is made even stronger than the original. This method is called oxy-acetylene welding.

Cutting with the oxy-fuel process is just the opposite from of welding. Oxy-fuel cutting uses acetylene and oxygen to preheat metal to red hot and then uses pure oxygen to burn away the preheated metal. Because this is achieved by oxidation, it is only effective on metals that are easily oxidized at this temperature. Such metals are mild steel and low allow steels. Oxy-fuel cutting can be used to cut thicknesses from 2/8″ to up to 12″.

Traditionally oxy-fuel processes are used for brazing, fusion welding, flame hardening, metalizing, soldering, stress relieving, cutting and bending. The primary uses today are welding, brazing and cutting.

### Chemistry of Oxy Acetylene Process

The most common fuel used in welding is acetylene. It has a two stage reaction; the first stage primary reaction involves the acetylene disassociating in the presence of oxygen to produce heat, carbon monoxide, and hydrogen gas.

$$2C_2H_2 + 2O_2 = 4CO + 2H_2 + Heat$$

A secondary reaction follows where the carbon monoxide and hydrogen combine with more oxygen to produce carbon dioxide and water vapor.

$$4CO + 2H_2 + 3O_2 = 4CO_2 + 2H_2O + Heat$$

When you combine equations (1) and (2) you will notice that about 5 parts of oxygen is necessary to consume 2 parts of acetylene

$$2C_2H_2 + 5O_2 = 4CO_2 + 2H_2O + Heat$$

Or we can say about 2.5 parts of oxygen is necessary to achieve complete combustion of acetylene. In operation, one part of oxygen is supplied through the torch and the remaining 1.5 parts is obtained from the surrounding air atmosphere (secondary reaction). When the secondary reaction does not burn all of the reactants from the primary reaction, the welding processes produces large amounts of carbon monoxide, and it often does. Because of the need for supplemental oxygen from the atmosphere, the acetylene oxygen flame cannot be used inside of pipes or structures subjected to oxygen depletion from gas welding. By varying the relative amounts of acetylene and oxygen, a welder can produce different flame atmospheres and temperatures as he requires.

## Oxy Fuel Welding Gases

Commercial fuel gases have one common property: they all require oxygen to support combustion. To be suitable for welding operations, a fuel gas, when burned with oxygen, must have the following:

a. High flame temperature

b. High rate of flame propagation

c. Adequate heat content

d. Minimum chemical reaction of the flame with base and filler metals

Among the commercially available fuel gases such as propane, liquefied petroleum gas (LPG), natural gas, propylene, hydrogen and MAPP gas, "Acetylene" most closely meets all the above requirements.

Acetylene is a hydrocarbon, just as are propane, methane, and virtually all the components which make up gasoline and fuel oils. However, it differs from those hydrocarbons in a way that its molecule is made up of two carbon atoms and two hydrogen atoms, the carbon atoms are joined by what chemists call a "triple bond". When acetylene reaches its kindling temperature; the bond breaks and releases energy. In other hydrocarbons, the breaking of the bonds between the carbon atoms absorbs energy. The triple bond is the reason that when acetylene and oxygen are mixed and ignited, the flame can reach the temperature of 5700° F to 6300° F, highest among commonly used gaseous fuels.

The other property of acetylene which you must remember is that acetylene/air mixtures can be ignited when they contain anywhere from 2.5 percent acetylene to 80 percent acetylene. Mixtures of methane (the principal component of natural gas) and air are flammable when they contain as little as 5% methane and not more than 15% methane.

The other two gases suitable for welding and cutting are:

a) MAPP

b) Hydrogen

MAPP gas is liquefied petroleum gas mixed with methylacetylene-propadiene (acetylene + propane). It has the storage and shipping characteristics of LPG and has a heat value a little less than acetylene. Because it can be shipped in small containers for sale at retail stores, it is used by hobbyist. MAPP gas can be used at much higher pressures than acetylene, sometimes up to 40 or 50 psi in high-volume oxy-fuel cutting torches which can cut up to 12 inch thick steel. A MAPP gas leak is easy to identify because of its particularly terrible odor.

Hydrogen produces low-temperature flame and is best for aluminum. Hydrogen flame is non-luminous, commonly used for underwater welding (can be used at higher pressure than acetylene).

Gases suitable for cutting but not welding:

a) Propane

b) Methane

c) LPG

Hydrocarbon gases, such as propane, butane, city gas, and natural gas, are not suitable for welding ferrous materials due to their oxidizing characteristics. Although propane has a very high number of BTUs per cubic feet in its outer cone, it does not burn as hot as acetylene in its inner cone and therefore not very useful for welding operations. However with a right torch (injector style), propane can make a faster and cleaner cutting and is much more useful for heating and bending applications than acetylene. In some instances, many nonferrous and ferrous metals can be braze welded with care taken in the adjustment of flare and the use of flux.

## Characteristics of Acetylene

1) Hydrocarbon - It burns in two stages with primary combustion zone (cone flame) and secondary combustion zone (envelope flame);

2) Acetylene is a colorless gas with a very distinctive, nauseating odor that is highly combustible when mixed with oxygen.

3) Acetylene gas is formed by the mixture calcium carbide and water and is composed of carbon and hydrogen having the chemical formula C2H2.

4) Neutral flame temperature of 5420°F at the end of the inner cone. This can be increased to a maximum of about 5800°F by increasing the oxygen quantity through the torch (oxidizing flame);

5) Combustible intensity is 12700 BTU/sec. /sq. ft. of cone area in a one to one mixture of acetylene and oxygen through the torch;

6) Net heat value of 1433 BTU per cubic foot;

### Advantages of Oxyacetylene Process

1) Does not require electricity;
2) The equipment is portable, easy to transport;
3) Welder has considerable control over the rate of heat input, the temperature of the weld zone, and the oxidizing or reducing potential of the welding atmosphere;
4) Oxyacetylene process is ideally suited to the welding of thin sheet, tubes, and small diameter pipe. It is also used for repair work, maintenance and in body shops;
5) Dissimilar metals can easily be joined;
6) Can also be used for preheating, cutting metal, case hardening, soldering and annealing.

## Shielded Metal Arc Welding

Shielded metal arc welding, commonly called stick, or covered electrode, welding, is a manual welding process that uses a welding power supply to create an electric an arc between a flux-covered consumable electrode and the workpiece to melt the metals at the welding point. The process uses the decomposition of the flux covering to generate a shielding gas and to provide fluxing elements to protect the molten weld-metal droplets and the weld pool. They can use either direct (DC) or alternating (AC) current, and consumable or non-consumable electrodes. The process uses the decomposition of the flux covering to generate a shielding gas and to provide fluxing elements to protect the molten weld-metal droplets and the weld pool.

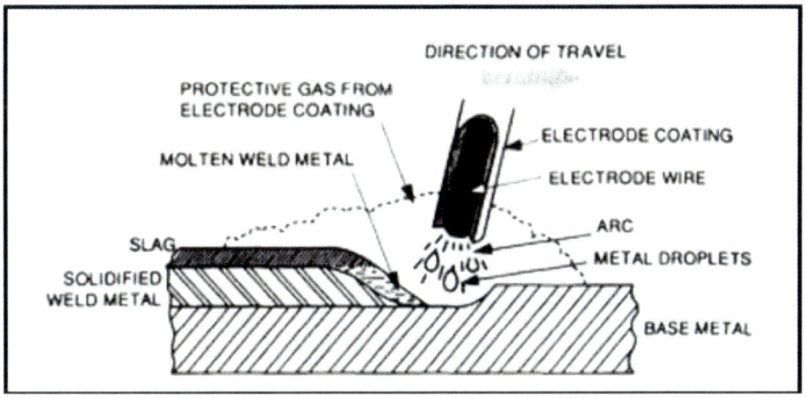


Figure: Shielded Metal Arc Welding

### Smaw Process

The SMAW process is the most widely used welding process. The arc is initiated by momentarily touching or "scratching" the electrode on the base metal. The resulting arc melts both the base metal and the tip of the welding electrode. The molten electrode metal and flux is transferred across the arc (by arc forces) to the base-metal pool, where the welding region is usually protected by some type of shielding gas, vapor, and/or slag.

## Striking and Establishing an ARC

A welding arc is maintained when the welding current is forced across a gap between the electrode tip and the base metal. A welder must be able to strike and establish the correct arc easily and quickly. There are two general methods of striking the arc:

1. Scratching

The scratching method is easier for beginners and when using an AC machine. The electrode is moved across the plate inclined at an angle, as you would strike a match. As the electrode scratches the plate an arc is struck. When the arc has formed, withdraw the electrode momentarily to form an excessively long arc, then return to normal arc length.

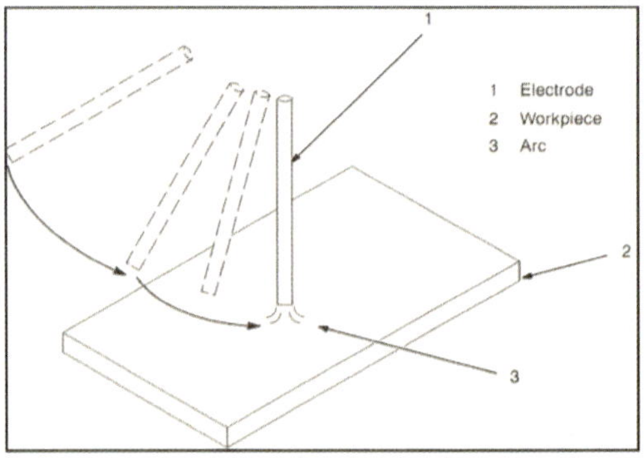


Figure: scratch method arc starting

2. Tapping

In the tapping method, the electrode is moved downward to the base metal in a vertical direction. As soon as it touches the metal it is withdrawn momentarily to form an excessively long arc, then returned to normal arc length.

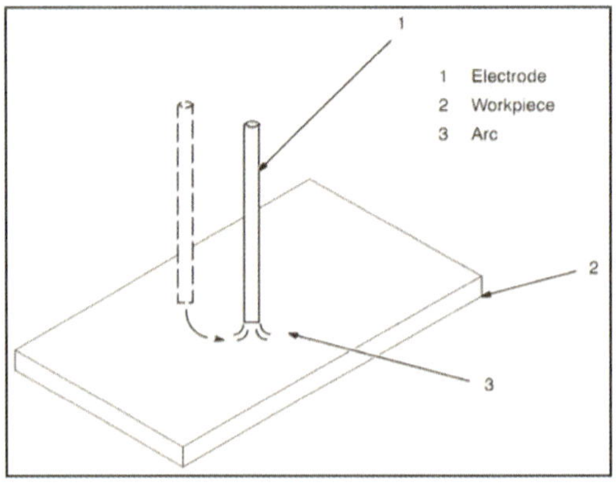


Figure: tapping method of arc starting

The principle difficulty encountered in striking the arc is freezing or when the electrode sticks or fuses to the work. This is caused by the current melting the electrode tip and sticking it to the cold base metal before it is withdrawn from contact the extra high current drawn by the short circuit will soon overheat an electrode and melt it or the flux, unless the circuit is broken. Giving the electrode holder a quick snap backward from the direction of travel will generally free the electrode. If it does not, it will be necessary to open the circuit by releasing the electrode from the holder.

## Welding Circuit

The circuit diagram for the SMAW process is shown in Figure. The equipment consists of a power source, electrode holder, and welding cables that connect the power source to the electrode holder and the workpiece. Alternating current (AC), or direct current, electrode negative (DCEN), or direct current, electrode positive (DCEP) can be used, depending on the electrode coating characteristics. The DCEN source is also called dc straight polarity, whereas the DCEP source is also called dc reverse polarity.

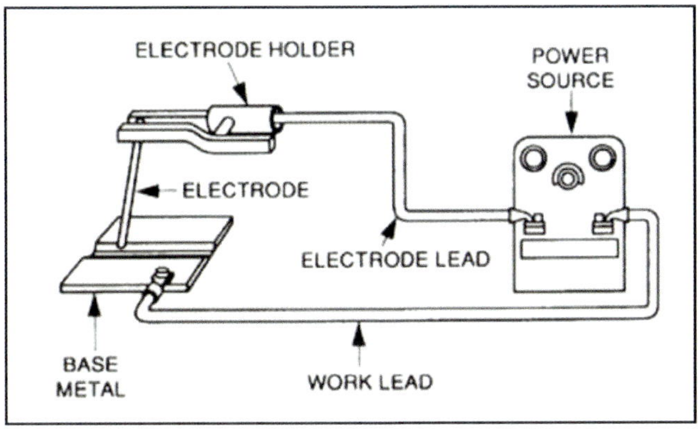


Figure: Welding circuit

The welding machine, or power source, is the crux of the SMAW process. Its primary purpose is to provide electrical power of the proper current and voltage to maintain a controllable and stable welding arc. Its output characteristics must be of the constant current (CC) type. SMAW electrodes operate within the range from 25 to 500 A. The electrode producer should suggest a narrow optimum range for each size and type of electrode. Operating arc voltage varies between 15 and 35 V.

## Electrode Holder

Electrode holder is held by the welder, firmly grips the electrode and transmits the welding current to it. Electrode holders are available in several designs, such as the pincher type and the collet, or twist, type, shown in figure. Each style has its proponents and the selection is usually a personal preference. Electrode holders are designated by their current capacity. Selection factors, such as the current rating, duty cycle, maximum electrode size, and cable size, are shown in table. The most lightweight holder that will accommodate the required electrode size is usually desired. All electrode holders should be fully insulated. Because they are used in proximity to the arc and are exposed to high heat, they will deteriorate rapidly. It is extremely important to maintain electrode holders to ensure that they retain their current-carrying efficiency, their insulating qualities, and their electrode gripping action. Manufacturers supply spare parts so that the holders can be rebuilt and maintained for safe and efficient.

## Electrodes

The electrodes used in the SMAW process have many different compositions of core wire and a wide variety of flux- covering types and weights. Standard electrode diameters of the core wire range from 1.6 to 8 mm. Electrode length usually ranges from 230 to 455 mm (9 to 18 in.); the shorter lengths are associated with the smaller diameter electrodes. A bare, uncoated end of the electrode (the grip end) is provided for making electrical contact in the electrode holder.

## Power Sources

Welding power sources are also known as welder, and welding machines. All welding machines are designed to change the high voltage alternating current main supply into a safe, low voltage, heavy-current supply, suitable for arc welding.

Arc welding machines produce either a constant current or a constant voltage. Manual metal arc welding process requires a constant current welding machine to minimize the changes in current whenever a welder changes the arc length.

Constant current welding machines are classified into three types. They are:

- 1-AC transformer
- 2-DC generator
- 3-AC /Dc Transformer-rectifier

Brief descriptions of these types are given below:

| Types of machine | Types of current available at the arc | Application |
|---|---|---|
| Transformer | AC | Used for both light and heavy steel fabrications and for TIG welding of aluminum |
| Transformer rectifier | AC and DC | As listed above plus light and heavy stainless steel fabrication. TIG welding of stainless steel, copper and nickel alloys using dc and aluminum using ac. |
| Motor generator | DC | Light and heavy fabrications made from low carbon steel and stainless steel |
| Engine generator | DC | As for motor generator but where main supply of electricity is not available |

## Applications

- The SMAW process has been replaced where possible because of greater productivity, higher quality, and, therefore, more cost-effective production. However, the ability of the SMAW process to achieve welds in areas of restricted accessibility means that it still finds considerable use in certain situations and applications.
- Heavy construction, such as shipbuilding, and welding "in the field," away from many support services that would provide shielding gas, cooling water, and other necessities, rely on the SMAW process to a great extent.
- Although the SMAW process finds wide application for welding virtually all steels and many of the nonferrous alloys, it is primarily used to join steels.
- This family of materials includes low-carbon or mild steels, low-alloy steels, high-strength

steels, quenched and tempered steels, high-alloy steels, stainless steels, and many of the cast irons.

- The SMAW process is also used to join nickel and its alloys and, to a lesser degree, copper and its alloys.
- It can be, but rarely is, used for welding aluminum.

## Advantages

- The SMAW process is the most widely used welding process.
- It is the simplest, in terms of equipment requirements
- The equipment investment is relatively small
- Welding electrodes are available for virtually all manufacturing, construction, or maintenance applications.
- Great flexibility of all the welding processes, because it can be used in all positions (flat, vertical, horizontal, and overhead)
- Virtually all base-metal thicknesses (1.6 mm, and greater), and in areas of limited accessibility, which is a very important capability
- Relatively high metal deposition rates
- Minimum of welding fume and of arc visibility (radiation)
- Suitable for welding carbon, low alloy and alloy steels.

## Limitations

- The most difficult in terms of welder training and skill-level requirements. Because the SMAW process is basically a manual process, the skill level of the welder is of paramount importance in obtaining an acceptable weld. The welder duty cycle is generally low, because of the built-in work break, which occurs after each electrode is consumed and requires replacement.
- Replacing the electrode when the arc is stopped (broken), the welder may "chip" or remove slag and clean it away from the starting and welding area with a wire brush to allow the proper deposition of the subsequent weld. This electrode replacement and cleaning operation occurs many times during the work day (about every two minutes, or the time it generally takes to consume an electrode). This stopping, chipping, wire brushing, and electrode replacement prevents the welder from attaining an operator factor, or duty cycle, that is much greater than 25%.
- Flat or horizontal position welding only
- Care required to preserve correct electrode alignment, as electrode
- Tip and weld pool are underneath solid flux cover

# Gas Tungsten Arc Welding

Gas tungsten arc welding is an electric arc welding process that produces an arc between a nonconsumable electrode and the work to be welded. The weld is shielded from the atmosphere by a shielding gas that forms an envelope around the weld area.

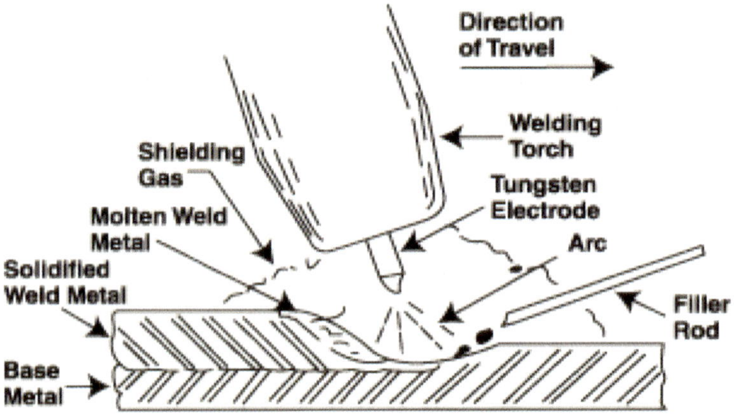


The GTAW process is versatile and can be used on ferrous and nonferrous metals. An arc is produced between a nonconsumable electrode and the work to be welded. The weld is shielded from the atmosphere by a shielding gas that forms an envelope around the weld area.

GTAW is versatile and can be used on ferrous and nonferrous metals and, depending on the base metal, in all welding positions. The process can be used to weld thin or thick materials with or without a filler metal.

When welding thinner materials, edge joints, and flange, filler metals are not used. For thicker materials, an externally fed filler wire is generally used. The type of filler metal wire to be used is based on the chemical analysis of the base metal. The size of the filler metal wire depends on the thickness of the base metal, which usually dictates the welding current.

The methods of operation for GTAW can be manual or automatic.

## Welding Procedure Variables and Joint Configurations

Welding procedure variables control the welding process and the quality of the welds produced. Joint configuration is determined by the design of the weldment, the metallurgical analysis, and by the process and procedure required by the weldment.

Welding variables are selected after the base metal, filler metal, and joint configuration have been selected. The fixed welding variables include the type of filler metal, electrode type and size, the type of current, and the type of shielding gas.

The adjustable variables control the shape of the weld by affecting things such as bead height, bead width, penetration, and weld integrity. The primary adjustable variables for GTAW are welding current, arc length, and travel speed.

Secondary variables also aid in controlling the welding process, but it is more difficult to calculate the extent of effect. The secondary variables include work and travel angle and the distance the electrode extends beyond the end of the cup.

## Tungsten Electrodes

The electrode material for GTAW is made from a tungsten alloy. Tungsten has one of the highest melting temperatures of any metal, about 6,170 degrees Fahrenheit (3,410 degrees Celsius).

The size of an electrode to be used is determined by the welding current required. Larger electrodes permit higher currents to be used. Smaller diameter electrodes may be used for welding thinner materials or while welding out of position.

Following is a list of various types of tungsten alloys used:

1. Pure tungstenis used on nonferrous metals, such as aluminum and magnesium, and is typically used with a balled-end preparation on alternating current (AC).

Figure: Pure tungsten is typically used with a balled-end preparation.

2. Thoriated tungstenis the most common type of tungsten electrode for use on carbon and stainless steel. It can be purchased with 1 or 2 percent thorium. The thoriated tungsten starts readily and maintains a stable arc. It has a greater resistance to contamination and will maintain a sharp point and will not break down as readily as pure tungsten.:
3. Zirconiated tungstenis typically used for welding with higher AC currents on nonferrous metals.

Preparing a point or using an electrode taper angle applies to thoriated tungsten. Thoriated tungsten electrodes are ground to a point to give better arc starting, with high frequency added. This provides the arc ignition and keeps the electrode from contacting the work. It also aids in stabilizing the arc.

The degree of taper affects the shape and amount of penetration of a weld. To reduce the number of times the electrode must be sharpened, the welder needs to develop the skill of not touching the tungsten to the work during the welding process. The recommended taper length is 21/2 to 3 times the diameter of the electrode.

## Shielding Gases

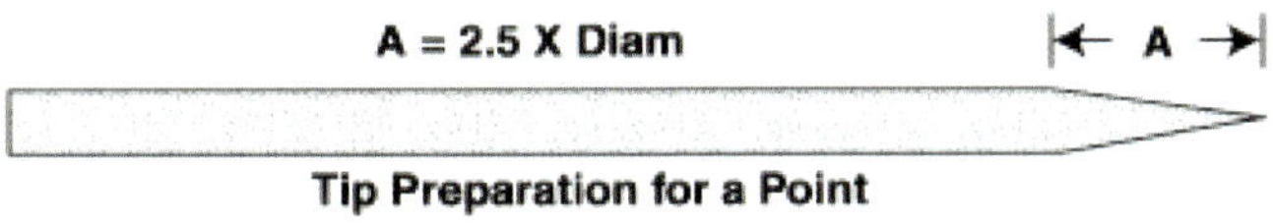


Figure : Proper electrode tip preparation is essential for achieving proper weld penetration.

Argon and helium are the two most commonly used shielding gases used for GTAW. The characteristics most desirable for shielding purposes are the chemical inertness of the gases and their ability to produce smooth arc action at high currents. Both gases are inert, causing an ionization effect in the welding arc. They protect the tungsten electrode and the molten weld pool from the atmosphere.

Gas purity affects a weld. Metals will withstand small amounts of impurities, but, for the best results, the percent of inert gas used should be at least 99.9 percent pure.

Argon is heavier than helium and may be supplied in liquid or gaseous form. Argon provides for good cleaning action. The flow rates are determined by the size of the tungsten and the gas cup diameter. Argon is suitable for welding similar and dissimilar metals and works well while welding in the vertical and overhead welding positions.

Helium is a lighter inert gas. It can be distributed as a liquid, but is used more often as a compressed gas. It leaves the weld area faster than argon, and higher flow rates are necessary when using it.

Helium produces a narrow but deep heat-affected zone (HAZ), which is good for welding on heavier metals. It is suitable for welding at high speeds and gives good coverage in vertical and overhead welding positions. It helps to increase the penetration, and when used as a back purge, it tends to flatten the pass of the weld bead. Helium is suitable for use on thicker nonferrous metals.

Argon and helium mixtures are used when welders need the control of the argon and the penetration of the helium. This mixture is not necessary when welding plain carbon steels.

Typical mixtures vary, depending on the application. It is often used for automatic welding applications.

Argon and hydrogen mixtures are often used for welding of stainless steel, INCONEL®, and MONEL®. This mixture should not be used when welding plain carbon steels. The typical mixture is a 95 percent argon and 5 percent hydrogen.

Nitrogen can also be used as a shielding gas, but is rarely used because of its higher current requirements. It is suitable for welding copper.

## Welding Current, Joint Design

The current depends primarily on the type of metal to be welded, the current levels required, and the availability of the machine that produces that type of welding current.

Direct current electrode positive (reverse polarity) is sometimes used to weld very thin nonferrous metals, and is also used for balling the tungsten electrode. Direct current electrode negative (straight polarity) is used most commonly to weld stainless steel and ferrous metals.

AC current, with the addition of high frequency, is most commonly used for welding some nonferrous metals such as aluminum and magnesium. It provides good cleaning action and gives moderate penetration.

## Weld Joint Design

The five basic types of joints are the butt joint, the corner joint, the edge joint, the lap joint, and the tee joint. Of the five types of joint designs, the butt and the tee joint are the most commonly used.

The strength of a weld joint is another factor contributing to weld joint design. Weld joints can be either partial or full penetration, depending on the strength required of the joint. Weld joint design or weldment configuration for GTAW is determined by the type of metal, configuration of the weldment, designated codes and specifications, and the metallurgical analysis. Several factors influence the joint design to be used, including the strength required, the welding position, the metal thickness, and how accessible the joint is to the welder.

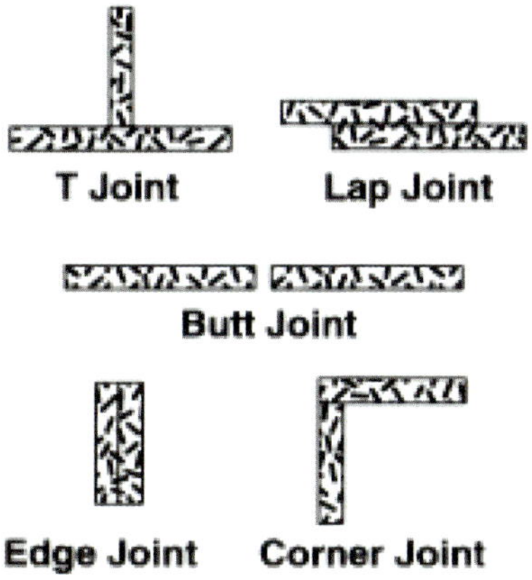


The purpose of any joint design is to produce a sound weld deposit with the desired properties as economically as possible. The edge and joint preparation are important because they will affect both the quality and the cost of welding.

## Preweld Preparations

Before using GTAW, several steps must be taken to prepare the electrode and the weld joint, fixture the weldment, set the variables, and preheat the base metal, if necessary. The amount of preparation depends on the size of the weldment, type of base material, fit-up, and the quality requirements.

Electrode Preparation: Electrode preparation depends on the type of electrode and the welding application. The tip may have a ground point or a ball end for welding with AC.

To prepare an electrode with a point, the grind marks should run parallel to the electrode.

To prepare a ball on the end of a tungsten, the power supply must be switched to DCEP (reverse polarity). Then, after starting the arc between the electrode and a piece of scrap metal or copper, it must be maintained at a moderate current level. The tip of the ball should be perfectly clean, shiny, and have a mirror-like finish.

Preparing the Weld Joint: When preparing the weld joint, several different methods can be used, including oxyfuel cutting, plasma cutting, shearing, machining, air carbon arc gouging, grinding, or chipping. Remember, preparing the weld joint properly will help produce a sound weldment and meet the requirements of quality standards for welding.

Cleaning: Cleaning the material to be welded is important. GTAW welds are often susceptible to contamination during welding. The surface to be welded must be free from oil, grease, paint, dirt, oxides, and other foreign material.

Aluminum has an oxide coating that, if not removed, will contaminate the weld area. Cleaning solutions, wire brushes, grinders, and abrasive blasting are some of the methods used to remove these contaminants.

Fixturing and Positioning: Fixturing and positioning will also affect the shape, size, and uniformity of a weld. Fixtures hold the weldment in place while controlling distortion, helping to locate and maintain parts in their position relative to the weldment.

When fixturing is employed, it can reduce the time for welding. Positioning will help move the weldment into a flat position to improve productivity for the welder.

Chill blocks, heat sinks, or backing bars may be used when welding some metals to prevent burn-through, reduce base material temperatures, or to minimize distortion.

Preheating: Depending on the alloying elements in the base material, the thickness of the steel, and the configuration of the joint, preheat is sometimes needed. The amount of preheat needed for a given application is usually obtained from the welding procedure. Several methods of controlling preheat temperature are available, including furnace heating, electric induction coils, oxy-fuel torches, and resistance heating blankets.

The preheat temperature can be measured using temperature sticks, crayon pellets, temperature indicators, thermocouples, thermistors, or infrared thermometers.

## Gas Metal Arc Welding

Gas metal arc welding, also called metal inert gas welding, is an arc welding process in which the heat for melting the metal is generated by an electric arc between a consumable electrode and the metal. The electrode is a solid wire, fed continuously through the arc into the weld pool, which ultimately becomes the filler metal at the weld. The wire type is selected to match the weld metal strength with the base metal strength. A mixture of inert gases, such as helium and argon, is streamed into the weld area to shield and protect the arc, weld pool, electrode and the base metal adjacent to the weld from interacting with the atmosphere. The welding parameters that are controlled to produce acceptable welds are arc current, arc voltage, wire feed speed, electrode travel speed, current density and preheat temperature. Preheating involves heating the base metal in the region surrounding the joint prior to welding and is often used to reduce the residual shrinkage stresses and increase the resistance to cracking in the weld area.

### MIG Welding Procedures

(1) The welding procedures for MIG welding are similar to those for other arc welding processes. Adequate fixturing and clamping of the work are required with adequate accessibility for the welding gun. Fixturing must hold the work rigid to minimize distortion from welding. It should be designed for easy loading and unloading. Good connection of the work lead (ground) to the workpiece or fixturing is required. Location of the connection is important, particularly when welding ferromagnetic materials such as steel. The best direction of welding is away from the work lead

connection. The position of the electrode with respect to the weld joint is important in order to obtain the desired joint penetration, fusion, and weld bead geometry. Electrode positions for automatic MIG welding are similar to those used with submerged arc welding.

(2) When complete joint penetration is required, some method of weld backing will help to control it. A backing strip, backing weld, or copper backing bar can be used. Backing strips and backing welds usually are left in place. Copper backing bars are removable.

(3) The assembly of the welding equipment should be done according to the manufacturer's directions. All gas and water connections should be tight; there should be no leaks. Aspiration of water or air into the shielding gas will result in erractic arc operation and contamination of the weld. Porosity may also occur.

(4) The gun nozzle size and the shielding gas flow rate should be set according to the recommended welding procedure for the material and joint design to be welded. Joint designs that require long nozzle-to-work distances will need higher gas flow rates than those used with normal nozzle-to-work distances. The gas nozzle should be of adequate size to provide good gas coverage of the weld area. When welding is done in confined areas or in the root of thick weld joints, small size nozzles are used.

(5) The gun contact tube and electrode feed drive rolls are selected for the particular electrode composition and diameter, as specified by the equipment manufacturer. The contact tube will wear with usage, and must be replaced periodically if good electrical contact with electrode is to be maintained and heating of the gun is to be minimized.

(6) Electrode extension is set by the distance between the tip of the contact tube and the gas nozzle opening. The extension used is related to the type of MIG welding, short circuiting or spray type transfer. It is important to keep the electrode extension (nozzle-to-work distance) as uniform as possible during welding. Therefore, depending on the application, the contact tube may be inside, flush with, or extending beyond the gas nozzle.

(7) The electrode feed rate and welding voltage are set to the recommended values for the electrode size and material. With a constant voltage power source, the welding current will be establish by the electrode feed rate. A trial bead weld should be made to establish proper voltage (arc length) and feed rate values.

Other variables, such as slope control, inductance, or both, should be adjusted to give good arc starting and smooth arc operation with minimum spatter.

The optimum settings will depend on the equipment design and controls, electrode material and size, shielding gas, weld joint design, base metal composition and thickness, welding position, and welding speed.

## Basic Concepts

MIG welding is operated in semiautomatic, machine, and automatic modes. It is utilized particularly in high production welding operations.

All commercially important metals such as carbon steel, stainless steel, aluminum, and copper can be welded with this process in all positions by choosing the appropriate shielding gas, electrode, and welding conditions.

## Equipment

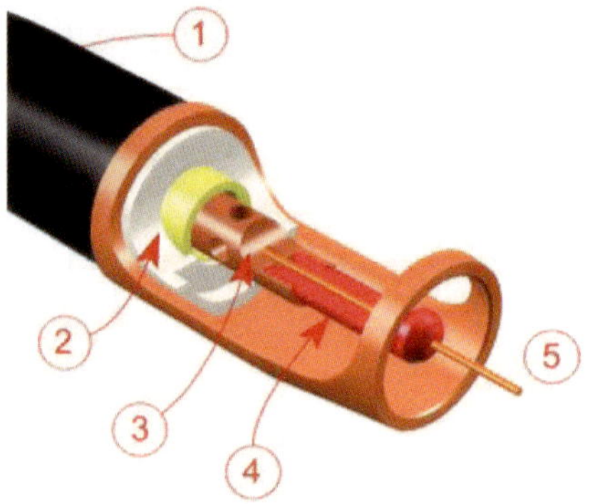


Gas metal arc welding equipment consists of a welding gun, a power supply, a shielding gas supply, and a wire-drive system which pulls the wire electrode from a spool and pushes it through a welding gun.

A source of cooling water may be required for the welding gun.

In passing through the gun, the wire becomes energized by contact with a copper contact tube, which transfers current from a power source to the arc.

While simple in principle, a system of accurate controls is employed to initiate and terminate the shielding gas and cooling water, operate the welding contractor, and control electrode feed speed as required.

The basic features of MIG welding equipment are shown in figure.

The MIG process is used for semiautomatic, machine, and automatic welding. Semiautomatic MIG welding is often referred to as manual welding.

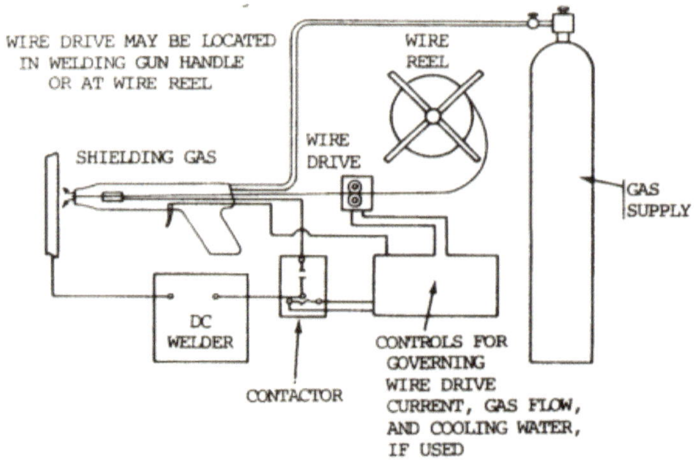


Developments during the 1950s and 1960s gave the process more versatility and as a result, it became a highly used industrial process.

Today, GMAW is commonly used in industries such as the automobile industry, where it is preferred for its versatility and speed.

Unlike welding processes that do not employ a shielding gas, such as shielded metal arc welding, it is rarely used outdoors or in other areas of air volatility.

A related Mig process, flux cored arc welding, often does not utilize a shielding gas, instead employing a hollow electrode wire that is filled with flux on the inside.

## Power Supply

Two types of power sources are used for MIG welding: constant current and constant voltage.

### (a) Constant Current Power Supply

With this type, the welding current is established by the appropriate setting on the power supply.

Arc length (voltage) is controlled by the automatic adjustment of the electrode feed rate.

This type of welding is best suited to large diameter electrodes and machine or automatic welding, where very rapid change of electrode feed rate is not required.

Most constant current power sources have a drooping volt-ampere output characteristic.

However, true constant current machines are available.

Constant current power sources are not normally selected for MIG welding because of the control needed for electrode feed speed. The systems are not self-regulating.

### (b) Constant Voltage Power Supply

The arc voltage is established by setting the output voltage on the power supply.

The power source will supply the necessary amperage to melt the welding electrode at the rate required to maintain the present voltage or relative arc length.

The speed of the electrode drive is used to control the average welding current.

This characteristic is generally preferred for the welding of all metals.

The use of this type of power supply in conjunction with a constant wire electrode feed results in a self-correcting arc length system.

Motor generator or dc rectifier power sources of either type may be used.

With a pulsed direct current power supply, the power source pulses the dc output from a low background value to a high peak value.

Because the average power is lower, pulsed welding current can be used to weld thinner sections than those that are practical with steady dc spray transfer.

## Manual and Automatic Welding Guns

Welding guns for MIG welding are available for manual manipulation (semiautomatic welding) and for machine or automatic welding.

Because the electrode is fed continuously, a welding gun must have a sliding electrical contact to transmit the welding current to the electrode.

The gun must also have a gas passage and a nozzle to direct the shielding gas around the arc and the molten weld pool.

Cooling is required to remove the heat generated within the gun and radiated from the welding arc and the molten weld metal.

Shielding gas, internal circulating water, or both, are used for cooling.

An electrical switch is needed to start and stop the welding current, the electrode feed system, and shielding gas flow.

## Semiautomatic MIG Welding Guns

Semiautomatic, hand-held guns are usually similar to a pistol in shape.

Sometimes they are shaped similar to an oxyacetylene torch, with electrode wire fed through the barrel or handle.

In some versions of the pistol design, where the most cooling is necessary, water is directed through passages in the gun to cool both the contact tube and the metal shielding gas nozzle.

The curved gun uses a curved current-carrying body at the front end, through which the shielding gas is brought to the nozzle.

This type of gun is designed for small diameter wires and is flexible and maneuverable.

It is suited for welding in tight, hard to reach corners and other confined places. Guns are equipped with metal nozzles of various internal diameters to ensure adequate gas shielding.

The orifice usually varies from approximately 3/8 to 7/8 in. (10 to 22 mm), depending upon welding requirements.

The nozzles are usually threaded to make replacement easier.

The conventional pistol type holder is also used for arc spot welding applications where filler metal is required.

The heavy nozzle of the holder is slotted to exhaust the gases away from the spot.

The pistol grip handle permits easy manual loading of the holder against the work.

The welding control is designed to regulate the flow of cooling water and the supply of shielding gas.

It is also designed to prevent the wire freezing to the weld by timing the weld over a preset interval.

A typical semiautomatic gas-cooled gun is shown in figure.

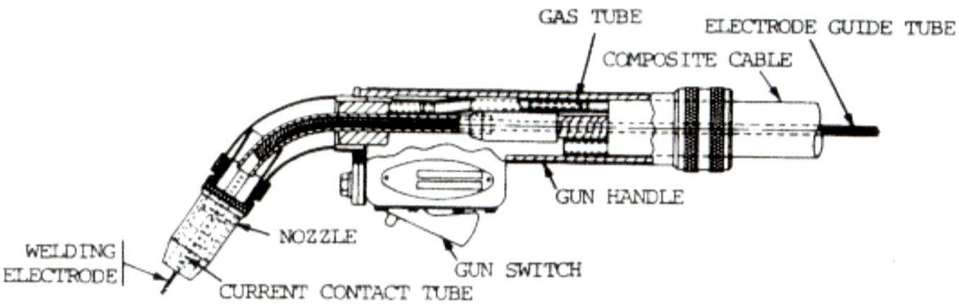


## Air Cooled Guns

Air-cooled guns are available for applications where water is not readily obtainable as a cooling medium. These guns are available for service up to 600 amperes, intermittent duty, with carbon dioxide shielding gas. However, they are usually limited to 200 amperes with argon or helium shielding. The holder is generally pistol-like and its operation is similar to the water-cooled type. Three general types of air-cooled guns are available:

1. A gun that has the electrode wire fed to it through a flexible conduit from a remote wire feeding mechanism. The conduit is generally in the 12 ft (3.7 m) length range due to the wire feeding limitations of a push-type system. Steel wires of 7/20 to 15/16 in. (8.9 to 23.8 mm) diameter and aluminum wires of 3/64 to 1/8 in. (1.19 to 3.18 mm) diameter can be fed with this arrangement.

2. A gun that has a self-contained wire feed mechanism and electrode wire supply. The wire supply is generally in the form of a 4 in. (102 mm) diameter, 1 to 2-1/2 lb (0.45 to 1.1 kg) spool. This type of gun employs a pull-type wire feed system, and it is not limited by a 12 ft (3.7 m) flexible conduit. Wire diameters of 3/10 to 15/32 in. (7.6 to 11.9 mm) are normally used with this type of gun.

3. A pull-type gun that has the electrode wire fed to it through a flexible conduit from a remote spool. This incorporates a self-contained wire feeding mechanism. It can also be used in a push-pull type feeding system. The system permits the use of flexible conduits in lengths up to 50 ft (15 m) or more from the remote wire feeder. Aluminum and steel electrodes with diameters of 3/10 to 5/8 in. (7.6 to 15.9 mm) can be used with these types of feed mechanisms.

## Water-cooled Guns

Water-cooled guns for manual MIG welding similar to gas-cooled types with the addition of water cooling ducts. The ducts circulate water around the contact tube and the gas nozzle. Water cooling permits the gun to operate continuously at rated capacity and at lower temperatures. Water-coded guns are used for applications requiring 200 to 750 amperes. The water in and out lines to the gun add weight and reduce maneuverability of the gun for welding.

## Air vs. Water Cooled Welding Guns

The selection of air- or water-cooled guns is based on the type of shielding gas, welding current range, materials, weld joint design, and existing shop practice. Air-cooled guns are heavier than water-cooled guns of the same welding current capacity. However, air-cooled guns are easier to manipulate to weld out-of-position and in confined areas.

## Arc Power and Polarity

The vast majority of MIG welding applications require the use of direct current reverse polarity (electrode positive). This type of electrical connection yields a stable arc, smooth metal transfer, relatively low spatter loss, and good weld bead characteristics for the entire range of welding currents used. Direct current straight polarity (electrode negative) is seldom used, since the arc can become unstable and erratic even though the electrode melting rate is higher than that achieved with dcrp (electrode positive). When employed, dcsp (electrode negative) is used in conjunction with a "buried" arc or short circuiting metal transfer. Penetration is lower with straight polarity than with reverse polarity direct current.

Alternating current has found no commercial acceptance with the MIG welding process for two reasons: the arc is extinguished during each half cycle as the current reduces to zero, and it may not reignite if the cathode cools sufficiently; and rectification of the reverse polarity cycle promotes the erratic arc operation.

## Metal Transfer

Filler metal can be transferred from the electrode to the work in two ways: when the electrode contacts the molten weld pool, thereby establishing a short circuit, which is known as short circuiting transfer (short circuiting arc welding); and when discrete drops are moved across the arc gap under the influence of gravity or electromagnetic forces. Drop transfer can be either globular or spray type.

Shape, size, direction of drops (axial or nonaxial), and type of transfer are determined by a number of factors. The factors having the most influence are:

- Magnitude and type of welding current.
- Current density.
- Electrode composition.
- Electrode extension.
- Shielding gas.
- Power supply characteristics.

Axially directed transfer refers to the movement of drops along a line that is a continuation of the longitudinal axis of the electrode. Nonaxially directed transfer refers to movement in any other direction.

## Short Circuiting Transfer

Short circuiting arc welding uses the lowest range of welding currents and electrode diameters associated with MIG welding. This type of transfer produces a small, fast-freezing weld pool that is generally suited for the joining of thin sections, out-of-position welding, and filling of large root openings. When weld heat input is extremely low, plate distortion is small. Metal is transferred from the electrode to the work only during a period when the electrode is in contact with the weld pool. There is no metal transfer across the arc gap.

The electrode contacts the molten weld pool at a steady rate in a range of 20 to over 200 times each second. As the wire touches the weld metal, the current increases. It would continue to increase if an arc did not form. The rate of current increase must be high enough to maintain a molten electrode tip until filler metal is transferred. It should not occur so fast that it causes spatter by disintegration of the transferring drop of filler metal. The rate of current increase is controlled by adjustment of the inductance in the power source. The value of inductance required depends on both the electrical resistance of the welding circuit and the temperature range of electrode melting. The open circuit voltage of the power source must be low enough so that an arc cannot continue under the existing welding conditions. A portion of the energy for arc maintenance is provided by the inductive storage of energy during the period of short circuiting.

As metal transfer only occurs during short circuiting, shielding gas has very little effect on this type of transfer. Spatter can occur. It is usually caused either by gas evolution or electromagnetic forces on the molten tip of the electrode.

## Globular Transfer

With a positive electrode (dcrp), globular transfer takes place when the current density is relatively low, regardless of the type of shielding gas. However, carbon dioxide ($CO_2$) shielding yields this type of transfer at all usable welding currents. Globular transfer is characterized by a drop size of greater diameter than that of the electrode.

Globular, axially directed transfer can be achieved in a substantially inert gas shield without spatter. The arc length must be long enough to assure detachment of the drop before it contacts the molten metal. However, the resulting weld is likely to be unacceptable because of lack of fusion, insufficient penetration, and excessive reinforcement.

Carbon dioxide shielding always yields nonaxially directed globular transfer. This is due to an electromagnetic repulsive force acting upon the bottom of the molten drops. Flow of electric current through the electrode generates several forces that act on the molten tip. The most important of these are pinch force and anode reaction force. The magnitude of the pinch force is a direct function of welding current and wire diameter, and is usually responsible for drop detachment. With $CO_2$ shielding, the wire electrode is melted by the arc heat conducted through the molten drop. The electrode tip is not enveloped by the arc plasma. The molten drop grows until it detaches by short circuiting or gravity.

## Spray Transfer

In a gas shield of at least 80 percent argon or helium, filler metal transfer changes from globular to spray type as welding current increases for a given size electrode. For all metals, the change takes place at a current value called the globular-to-spray transition current.

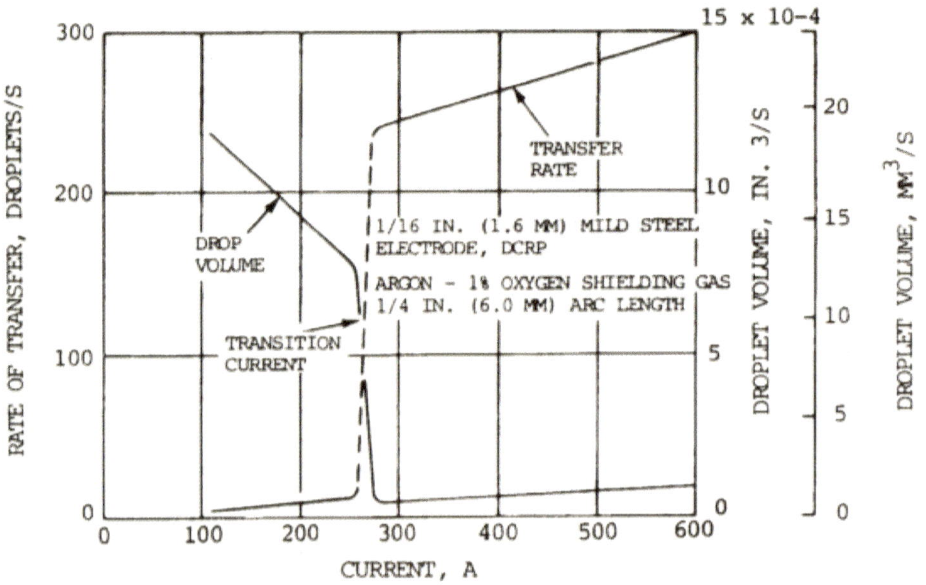


Variation in volumes and transfer rate of drops with welding current (steel electrode)

Spray type transfer has a typical fine arc column and pointed wire tip associated with it. Molten filler metal transfers across the arc as fine droplets. The droplet diameter is equal to or less than the electrode diameter. The metal spray is axially directed. The reduction in droplet size is also accompanied by an increase in the rate of droplet detachment, as illustrated in figure. Metal transfer rate may range from less than 100 to several hundred droplets per second as the electrode feed rate increases from approximately 100 to 800 in./min (42 to 339 mm/s).

## Advantages of MIG Welding

(1) The major advantage of gas metal-arc welding is that high quality welds can be produced much faster than with SMAW or TIG welding.

(2) Since a flux is not used, there is no chance for the entrapment of slag in the weld metal.

(3) The gas shield protects the arc so that there is very little loss of alloying elements as the metal transfers across the arc. Only minor weld spatter is produced, and it is easily removed.

(4) This process is versatile and can be used with a wide variety of metals and alloys, including aluminum, copper, magnesium, nickel, and many of their alloys, as well as iron and most of its alloys. The process can be operated in several ways, including semi- and fully automatic. MIG welding is widely used by many industries for welding a broad variety of materials, parts, and structures.

## Disadvantages of MIG Welding

(1) The major disadvantage of this process is that it cannot be used in the vertical or overhead welding positions due to the high heat input and the fluidity of the weld puddle.

(2) The equipment is complex compared to equipment used for the shielded metal-arc welding process.

# Flux-cored Arc Welding

Flux-cored arc welding utilizes a continuously-fed tubular electrode, electrical power to melt the electrode, and may or may not use shielding gas from an externally supplied source when depositing material in the weld joint. The flux-cored electrode is a composite tubular filler metal consisting of a metal sheath with a core of mineral compounds and powdered metals. The resulting weld bead is covered by a residual slag (melted and solidified flux) which helps to protect and shape the finished weld, and is generally easily removed upon completion of the weld. Flux Cored Welding combines the high productivity of MIG welding using a solid electrode with the ability to weld on more contaminated base material. When compared with MIG welding, higher deposition rates are possible, especially when welding out of position.

## Flux Cored Arc Welding Production

Flux cored arc welding is the most productive of the manual welding processes! When comparing MIG welding to flux core arc welding, there is a huge gap in production, in the amount of weld per hour. A MIG welder can typically produce 5 to 8 pounds of weld per hour, versus a FCAW welder packing 25 plus pounds of weld per hour. On top of that flux core welding can weld 1/2″ plates in a single pass with full penetration on both sides. Flux core arc welding for this reason is primarily used in the shipbuilding industry. Ships are made of heavy plate, and have endless amounts of welding that needs to be done. Flux core welding produces high quality welds, fast, and even when in windy conditions.

## Weldability of Metals

Welding with flux cored electrodes has some serious cons when it comes to the weldability of metals. So far, FCAW has been perfected on most carbon steels, cast iron, nickel based alloys, and some stainless steels. Unfortunately most non ferrous exotic metals cannot be welded and that includes aluminum. On the upside for most hobbyists the flux core electrode may be an excellent choice for general garage work, because if used in a MIG welder there is no shielding gas required on some electrodes.

## Working of Flux Cored Arc Welding

Flux cored arc welding just like MIG welding requires three main ingredients, electricity, filler metal, and a form of shielding from the air. Just like MIG welding, flux core welding works by feeding an electrode continuously to the joint. First the welder squeezes the trigger, and then the wire feeder begins to feed the electrode to the joint, at the same time the electrode gets electrically charged. Once the electrode hits the metal joint, the electricity short circuits, and heats up the electrode till the electrode begins to melt. Once the electrode begins to melt, the metal also starts to melt, and then both of them start to create a puddle. This puddle at the same time melts the flux core, creating a shield from air, and at the same time produces a slag that protects the weld from contamination.

## FCAW Self-shielding vs. Dual Shield

Flux cored arc welding comes in two types of shielding. The first difference is in the electrode itself, it is a tubular wire with a shielding powder in the center. In technical terms this is called "self-shielding" or sometimes branded "inner shield". The second is the same type of electrode, but another ingredient is added. A bottle of gas is used in addition to the flux core shielding. The technical term for this is "dual shield". In the case of dual shielding, you have a powder flux in the center of the electrode and an external shielding gas protecting the weld area.

## FCAW Voltage Type – Welding Polarity – Power Supply

A flux cored welding power supply is also a MIG welding power supply, they are the same machine. That is a "Constant Voltage Power Supply". Constant voltage power supplies keep the voltage near, or at the same level. Unlike a TIG, or Stick welder, that keeps the amperage consistent. In the flux cored welder the amperage is changed with the wire feed speed. The faster the wire feeds, the more contact the electrode has, producing more amperage, and heat.

The voltage type used is D/C direct current like the type current produced by a battery. The polarity used in industrial flux core arc welding is typically D/C electrode (+) positive. This means that the handle is the positive side of the circuit, or the electricity flows from the metal to the welding handle. This is typical when larger electrodes are used. When welding with smaller electrodes and sheet metals, the polarity is changed to D/C electrode (-) negative.

The main difference between FCAW, and a MIG welder's are, flux cored arc welding power supplies are available with, much, more, power! Basically they are an extremely powerful MIG welder! Some flux core arc welders come with the capabilities of running over an extremely hot, 1000 plus amps. That is where they leave MIG welding in the dust for production.

## Shielding Gases for FCAW

In the case of dual shielding being used with a flux cored electrode the choices of shielding gasses are limited.

The choices are as follows:

$CO_2$ – Carbon dioxide

Ar – Argon

$CO_2$ / Ar – A mixture of the two

Ar / Ox – A mixture of the two

$CO_2$ by itself produces the deepest penetrating weld but has some drawbacks. The mechanical properties of the weld are not the best due to fact the flux in the wire reacts with the shielding gas. Others drawbacks are, it produces a lot of spatter, and the arc is stiff and not as stable as it can be.

Argon by itself will also weld with a flux cored electrode, but just like $CO_2$, it reacts not favorable with the flux. Both Argon and Carbon dioxide can make a decent looking weld if used by themselves. What the weld looks like versus the actually quality of the weld are two different stories.

The most common gases used for dual shield FCAW are a mixture of Carbon Dioxide and Argon or Argon and Oxygen. The most popular is C25 / 25% Carbon Dioxide and 75% Argon. This gas produces a stable arc, less spatter, and allows more of a spray transfer of metal.

## FCAW Electrode Types

The electrodes used for flux cored welding are almost visually the same as a MIG welding electrode. The difference is that flux cored electrodes are tubular, or a hollow tube with flux in the center. MIG welding electrodes are solid metal.

Flux cored electrodes come in standard sizes. Some are the same size as most MIG welding electrodes but others are comparable to the thickness of a stick welding electrode.

Here are some of the more popular sizes for standard industrial applications:

035, 045, 052, 1/16.

As with most electrodes there is a standard classification code or designation code, on the spool they come on. To understand the classifications better it is important to know some basics about where the classification codes are different.

A somewhat common flux cored welding electrode is the "E71T – 1". As with all electrodes the numbers and letters all mean something. There identifications are as follows:

E – Stands for electrode.

- 7 – Stands for the minimum amount of tensile strength. In this case it is 70,000 lbs of tensile strength per square inch of weld. The way this number is figured is by adding four zeros to the number.
- 1– Stands for the position that this electrode can be welded in. There are only two designations and they are "0" for flat and horizontal welding, then there is "1" for all position welding.
- T – Stands for a tubular electrode. When "T" is used it is always assumed it is a flux cored electrode.
- 1 – The last is the shielding flux type designations.

As a note with all flux cored electrodes they need to be stored in a dry place. Otherwise the may pick up moisture and this will cause major weld defects.

## Flux Cored Welding Transfer Types

When welding with a flux cored electrode there are two metal transfer types used. The transfer types are spray transfer and globular. Spray transfer is the most commonly used. Just like the name states the metal from the electrode gets heated up to the point that it literally sprays the filler metal to the joint. Globular transfer heats up the electrode hot enough for globs of metal to drip off of the electrode to the weld joint. What separates the two transfer types are, voltage settings, wire speed, and the gases used, if any.

### Flux Cored Arc Welding Joint Preparation

Joint preparation for flux core is not as critical as with MIG welding. FCAW can typically burn through mill scale and minor rust. In many cases when the metal is cut with a torch, it can be welded as-is, with no additional cleaning. For the shipbuilding industry this is a huge savings in labor cost. In addition to easy joint preparation, beveled groove joints can be narrower for metals ½ inch or thinner, and they can be welded in a single pass with full penetration on both sides.

### Ceramic Backing Tape

Commonly in the ship building industry many joints are welded from a single side using a ceramic backing tape. The ceramic backing tape is like a mold to pour metal in, but in this case the electrode will fill that mold. When ceramic backing tape is used it allows for full joint preparation and outstanding weld quality. This in return gives full control of the shape and penetration of the back side of the weld. Once the weld is finished, the ceramic tape is simply peeled off, and thrown away.

## Submerged Arc Welding

Submerged arc welding is an arc welding process in which heat is generated by an arc which is produced between bare consumable electrode wire and the workpiece. The arc and the weld zone are completely covered under a blanket of granular, fusible flux which melts and provides protection to the weld pool from the atmospheric gases.

The molten flux surrounds the arc thus protecting arc from the atmospheric gases. The molten flux flows down continuously and fresh flux melts around the arc. The molten flux reacts with the molten metal forming slag and improves its properties and later floats on the molten/solidifying metal to protect it from atmospheric gas contamination and retards cooling rate. Process of submerged arc welding is illustrated in figure.

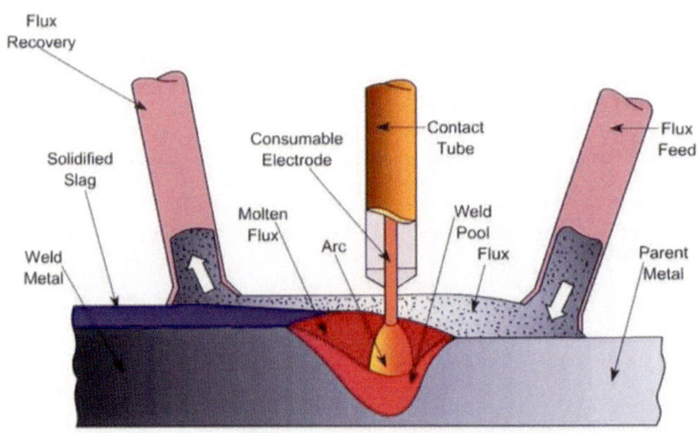


Figure : Process of Submerged Arc Welding

Extremely high welding currents can be used without the danger of spatter and atmospheric contamination giving deep penetration with high welding speeds. A proper selection of flux-wire combination can produce welds of very high quality. This makes the process very suitable for the welding of high strength steel at welding speeds much higher than conventional manual metal arc welding. It is found that the desired composition of the weld metal can be more economically obtained through adding alloying elements in the flux and using a relatively unalloyed wire as compared with welding with alloyed wire and ordinary flux.

A continuous consumable wire electrode is fed from a coil through contact tube which is connected to one terminal of power source. Wires in the range 1 – 5 mm diameters are usually employed and with wires at the lower end of this range (upto 2.4 mm) constant-potential DC power source can be used allowing arc length control by the self-adjusting effect. For higher diameter electrodes constant current DC source is used.

Submerged arc welding head may be mounted on self-propelled tractors carrying a flux hopper and the coiled electrode. A suction device may also be carried to recover the unused flux for reuse.

Since the end of the electrode and the welding zone are completely covered at all times during the actual welding operation, the weld is made without the sparks, spatter, smoke or flash commonly observed in other arc welding processes.

Power source requirement may be DC or AC. Normally electrode is connected to positive terminal of DC power source. Sometime depending on the nature of flux AC can be used with single electrode wire or with multiple electrodes where one electrode may be connected to DC and other to AC if independent power sources are to be used.

Electrode wires and fluxes are two major consumables. Wires of structural steel are coated with copper to protect it from atmospheric corrosion and increasing its current carrying capacity while stainless steel wires are not coated with copper.

Flux in submerged arc welding performs more or less the similar functions as the electrode coating in the case of MMA welding, except from generation of shielding gas. However, these fluxes perform additional function of pickup or loss of alloying elements through gas metal and slag metal reactions as the molten flux gets sufficient time to react with molten metal and performs above reactions and then forming slag. Some fluxes require baking before use, to remove moisture which might have been absorbed during storage. Such fluxes should be baked as per manufacturer's recommendations or at 250–300°C for 1 - 2 hours duration before use.

Fluxes are fused or agglomerated consisting of MnO, $SiO_2$, CaO, MgO, $Al_2O_3$, $TiO_2$, FeO, and $CaF_2$ and sodium/potassium silicate. Particular flux may consist of some of these constituents and other may not be present. Depending upon the flux constituents the base of flux is decided. Also the basicity index of flux is decided on the flux constituents. The ratio of contents of all basic oxides to all acidic oxides in some proportion is called basicity index of a flux. CaO, MgO, BaO, $CaF_2$, $Na_2O$, $K_2O$, MnO are basic constituents while $SiO_2$, $TiO_2$, $Al_2O_3$ are considered to be acidic constituents.

When welding with low basicity index fluxes, better current carrying capacity, slag detachability and bead appearance are achieved while mechanical properties and crack resistance of the weld metal are poor. High basicity fluxes produce weld metal with excellent mechanical properties and resistance to cracking, however, bead appearance and current carrying capacity are poor.

Electrode wire size, welding voltage, current and speed are four most important welding variables apart from flux. Welding current is the most influential variable as it controls electrode melting rate, depth of penetration and the amount of base metal fused. However, very high current shall lead to too much penetration resulting into burn through in the metal being joined, excessive reinforcement and increased weld shrinkage and, therefore, large amount of distortion. On the other hand low current shall lead to insufficient penetration, lack of fusion and unstable arc.

Welding voltage has nominal effect on the electrode wire melting rate but high voltage leads to flatter and wider bead, increased flux consumption and resistance to porosity caused by rust or scale and helps bridge gap when fitup is poor. Lower voltage produces resistance to arc blow but high narrow bead with poor slag removal. Welding voltages employed vary from 22 to 35 V.

If the welding speed is increased, power or heat input per unit length of weld is decreased, less welding material is applied per unit length of weld, and consequently less weld reinforcement results and penetration decreases. Travel speed is used primarily to control bead size and penetration. It is interdependent with current.

Excessive high travel speed decreases wetting action, increases tendency for undercut, arc blow, porosity and uneven bead shapes while slower travel speed reduces the tendency to porosity and slag inclusion.

The electrode size principally affects the depth of penetration for fixed current. Small wires are generally used in semiautomatic equipment to provide flexibility to the welding gun. The small wires are also used in multiple electrodes, parallel wire setups.

The larger electrodes are generally used to take advantage of higher currents and consequently higher deposition rates. Where poor fitup is countered a larger electrode is capable of bridging gaps better than smaller ones.

Variations in submerged arc welding may be single electrode wire and multiple electrode wires i.e. multiple arcs.

Multiple arcs are used to increase deposition rates and to direct the arc blow in order to provide an increase in welding speed. Multiple arcs may also reduce the solidification rate and porosity in the weld metal. Multiple arcs may be used either with a single power source or with separate power sources for each electrode.

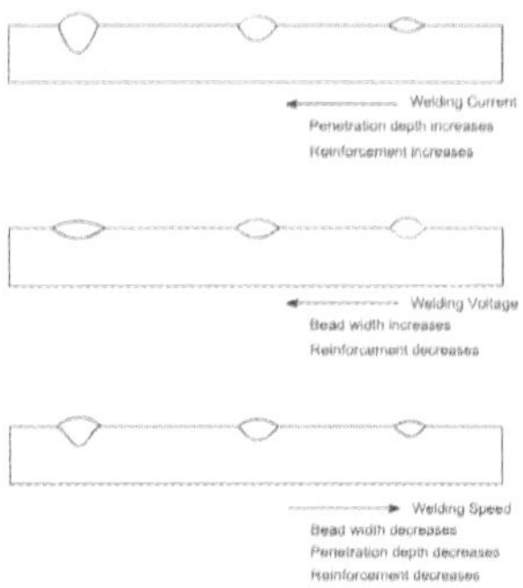


Figure: Influence of Welding Parameters on Bead Shape.

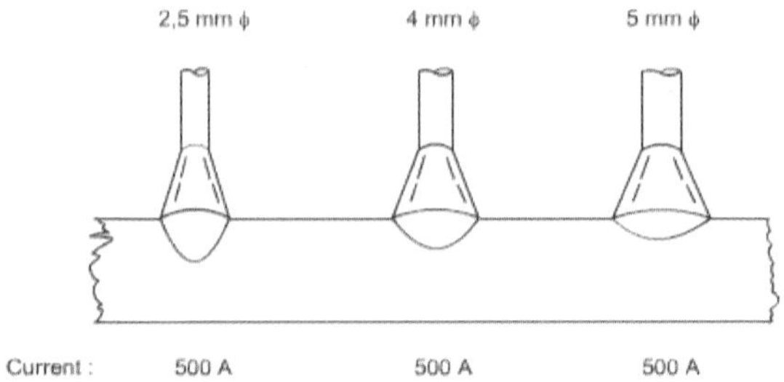


Figure: Influence of Electrode Size on Weld Bead.

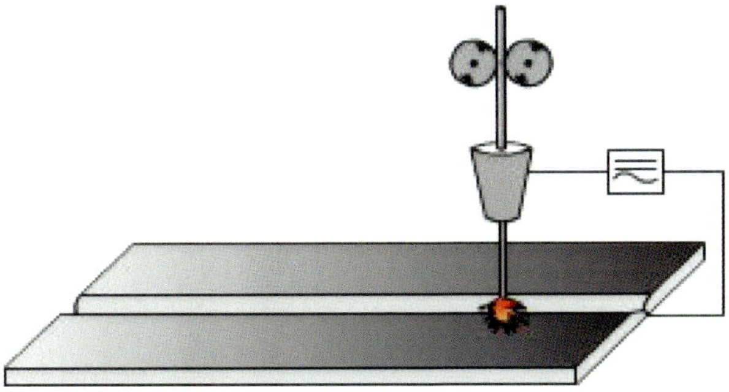

Figure : Conventional Single Electrode SAW.

Submerged arc welding process has high deposition rate with high depth of penetration. This is continuous welding process with no interruptions as electrode wire is supplied through coil on a spool. Welding is carried out without sparks, smoke or spatter. Weld bead is very clean and smooth. Welds produced are of high quality with good mechanical and metallurgical characteristics.

As the arc is not visible, being covered with the layer of slag, so it necessitates accurate guidance of the welding head on the weld groove, failing which an improper fusion will result. Further, process can be used only in flat or HV positions. Plates of lesser thickness (less than 5 mm) can not be welded because of danger of burn through which may occur. Circumferential welds cannot be made in small diameter components because the flux falls away.

Submerged arc welding is mainly being used for different grades of steels. It is widely being used in shipbuilding, offshore, structural and pressure vessel industries. General fabrication such as fabrication of pipes, penstocks, LPG cylinders, bridge girders and other structures are produced by SA welding. Surfacing for reclamation of worn out parts or for deposition of wear or corrosion resistant layers or for hardfacing layers also employ submerged arc process.

## Electroslag Welding

Electroslag welding is a welding process, in which the heat is generated by an electric current passing between the consumable electrode (filler metal) and the work piece through a molten slag covering the weld surface.

Prior to welding the gap between the two work pieces is filled with a welding flux. Electroslag Welding is initiated by an arc between the electrode and the work piece. Heat, generated by the arc, melts the fluxing powder and forms molten slag. The slag, having low electric conductivity, is maintained in liquid state due to heat produced by the electric current.

The slag reaches a temperature of about 3500°F (1930°C). This temperature is sufficient for melting the consumable electrode and work piece edges. Metal droplets fall to the weld pool and join the work pieces.

Electroslag welding is used mainly for steels.

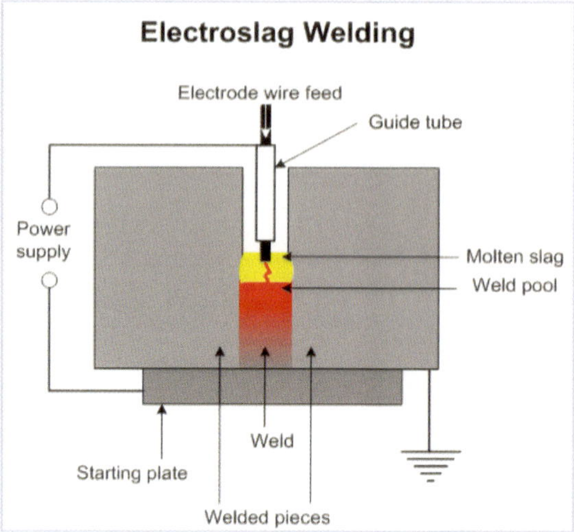


## Principle

Electroslag welding process works on the principle of heat generation due to arc and electric resistance. Arc is made between welding electrode and the workpiece which starts melting the filler metal to fill the filler cavity. Now the heat is produced due to electric resistance when the current passes through the surface. The heat further starts melting the filler metal which is continuously fed from the roller. This filler wire melts and fills the weld and makes it a strong joint.

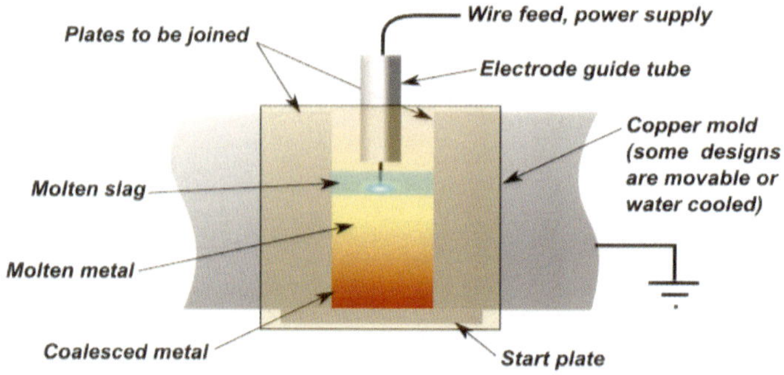


Electrodes: Generally two types of electrodes that are solid and metal-cored are used. Though solid electrodes are more popular than the metal-cored electrodes.

Flux: Flux is the most important consumable material of Electroslag Welding. In its molten state it transforms the electrical energy into heat energy which helps in melting the electrode wire and the base metal to form a weld joint. It is also required to protect the molten weld metal from the atmosphere and to ensure stability. The flux in its molten state is required to conduct electricity but at the same time it should offer sufficient resistance to its flow for generating enough heat to do welding.

Arcing results when the resistance is less than the minimal required quantity. The slag must also have minimum viscosity so that it does not be too thick to stop good circulation and nor too thin to result in excessive leakage.

Electrode Guide Tube: It is used to guide the electrode wire at desired position where the welding is to be done.

## Working of Electroslag Welding

It works as follows:

- Firstly current flows from welding electrode to base plate. This establishes an arc between electrode and base plate which heats the flux. This heat which is produced during arc formation results in melting the filler metal and deposits into the weld cavity.
- Now the cooled copper shoe starts its function of solidifying the filler metal into weld cavity. This is done to avoid flowing of weld metal outside.
- As the filler metal solidifies into weld cavity, the current flows through it. Then it generates heat due to electric resistance. This heat is further used to continue the melting of the filler metal into weld Cavity. Which means that heat is regenerated, which results in less waste of heat or energy.
- Roller arrangement continuously provides the filler metal.
- During welding of the metals both copper shoe and feed mechanism moves upward until the whole cavity is formed.
- This will create a strong joint in single one pass. The single or multi-pass weld is used according to plate thickness.

## Advantages

- Cooling rate is very low so there is no problem of cold cracking.
- There is no problem of slag inclusion or porosity.
- The process is semi-automatic and faster.
- Heavier section can be welded in single pass.
- High productivity can be achieved.
- Low cost for joint preparation.

## Disadvantages

- Too high heat input to base.
- High temperature of welding needs cooling arrangement.
- Slow rate of cooling give columnar grain in weld.
- When the heat input is very high then the weld quality can be poor, including low toughness caused by the coarse grains in the fusion zone and the heat-affected zone.
- In Electroslag welding, it has some tendency towards hot cracking and notch sensitivity in the heat affected zone.

- It is restricted to vertical position welding, because of large molten metal pools and slag.
- It is difficult to close cylindrical welds
- It tends to produce large grain sizes..
- If you are creating joints below 60 mm than Submerged Arc Welding (SAW) is more economical than electroslag welding process.

### Application

- It is used in heavy industries where plate thickness is up to 80 mm to be joined.
- Welding of thick walled large diameter pipes is done by this welding process.
- Welding of storage tanks is done by it.
- It is used to construct big and thick parts of ships.

## Resistance Welding

Resistance welding is a welding technology widely used in manufacturing industry for joining metal sheets and components. The weld is made by conducting a strong current through the metal combination to heat up and finally melt the metals at localized point(s) predetermined by the design of the electrodes and/or the workpieces to be welded. A force is always applied before, during and after the application of current to confine the contact area at the weld interfaces and, in some applications, to forge the workpieces.

### Resistance Welding Processes

Depending on the shape of the workpieces and the form of the electrodes, resistance welding processes can be classified into several variants among which the most commonly used are spot welding, projection welding, seam welding and butt welding. More details are described below:

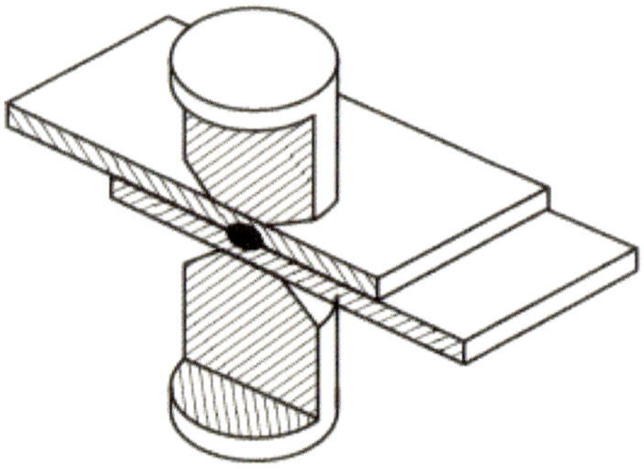

### Resistance Spot Welding

Spot welding is a resistance welding process for joining metal sheets by directly applying opposing forces with electrodes with pointed tips. The current and the heat generation are localized by the form of the electrodes. The weld nugget size is usually defined by the electrode tip contact area.

Spot welding is the predominant joining process in automotive industry for assembling the automobile bodies and large components. It is also widely used for manufacturing of furniture and domestic equipment etc.

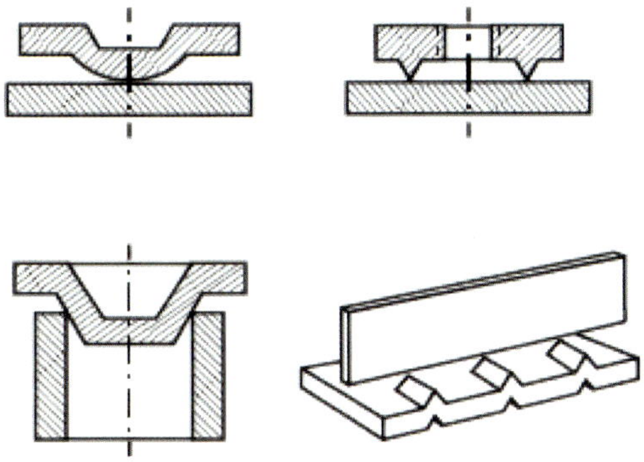

## Resistance Projection Welding

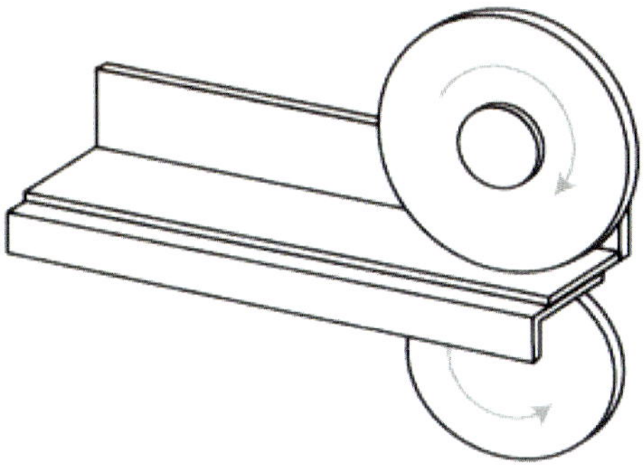

Projection welding is a resistance welding process for joining metal components or sheets with embossments by directly applying opposing forces with electrodes specially designed to fit the shapes of the workpieces. The current and the heat generation are localized by the shape of the workpieces either with their natural shape or with specially designed projection. Large deformation or collapse will occur in the projection part of the workpieces implying high process/machine dynamics.

Projection welding is widely used in electrical, electronics, automotive and construction industries, and manufacturing of sensors, valves and pumps etc.

## Resistance Seam Welding

Seam welding is a resistance welding process for joining metal sheets in continuous, often leak tight, seam joints by directly applying opposing forces with electrodes consisting of rotary wheels. The current and the heat generation are localized by the peripheral shapes of the electrode wheels.

Seam welding is mostly applied in manufacturing of containers, radiators and heat exchangers etc.

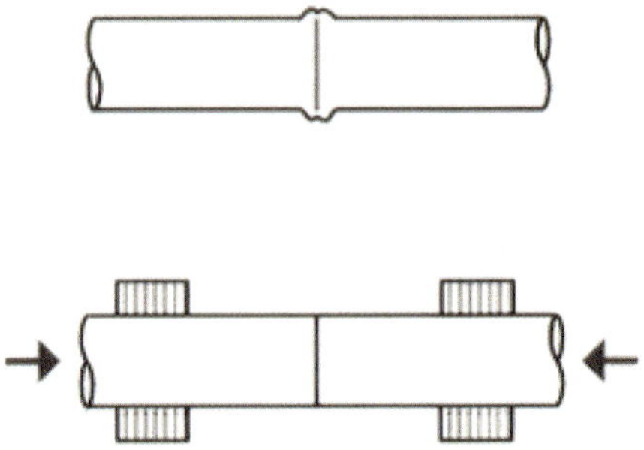

### Resistance Butt Welding

Butt welding is a resistance welding process for joining thick metal plates or bars at the ends by directly applying opposing forces with electrodes clamping the workpieces. A forging operation is applied after the workpieces are heated up. Often no melt occurs, thus a solid state weld can be obtained.

Butt welding is applied in manufacturing of wheel rims, wire joints and railway track joints etc.

### Single-sided (One-sided) Resistance Welding

Is a special resistance welding process where a spot weld is made with only one electrode accessing from one side to the weld zone with or without a backing plate from the other side. Low weld force is usually used, which limits the single-sided (one-sided) spot welding to joining of relatively thin sheets. It may be useful for welding components with limitation of electrode access from both sides.

### Resistance Weld Bonding

It is a combined joining process with adhesive bonding and resistance welding. The adhesive is applied to the faying surfaces of sheets to be welded, and subsequently resistance spot weld is made through the sheets before curing of the adhesive. The joint can have good strength from the spot welding and good stiffness from the adhesive bonding.

### Cross Wire Welding

It is a resistance welding process for joining bars or wires in cross joints by directly applying opposing forces with usually flat electrodes. The current and the heat generation are localized at the contact points of the crossed bars or wires. Cross wire welding is widely used in construction and electrical industry as well as for manufacturing of metal wire nets and shopping trolleys etc.

### Indirect Spot Welding

It is a special resistance welding process where a single spot weld is made with one electrode directly connecting to the weld zone, while the other electrode is offset at a distance, but still conducts the current along the workpiece.

### Series Spot Welding

It is a special resistance welding process where two spot welds are made at the same time with two electrodes offset at a distance but still conducting the current along the workpieces between the two welds.

## Micro Resistance Welding

It refers to the resistance welding processes for joining micro or miniaturized components, which in principle can be any of the above mentioned process variants but in a micro scale.

## Parameters in Resistance Welding

The principle of resistance welding is the Joule heating law where the heat Q is generated depending on three basic factors as expressed in the following formula:

$$Q = I^2 R\, t$$

where I is the current passing through the metal combination, R is the resistance of the base metals and the contact interfaces, and t is the duration/time of the current flow.

The principle seems simple. However, when it runs in an actual welding process, there are numerous parameters, some researchers had identified more than 100, to influence the results of a resistance welding. In order to have a systematic understanding of the resistance welding technology, we have carried out a lot of experimental tests and summarized the most influential parameters into the following eight types:

## 1) Welding Current

The welding current is the most important parameter in resistance welding which determines the heat generation by a power of square as shown in the formula. The size of the weld nugget increases rapidly with increasing welding current, but too high current will result in expulsions and electrode deteriorations. The figure below shows the typical types of the welding current applied in resistance welding including the single phase alternating current (AC) that is still the most used in production, the three phase direct current (DC), the condensator discharge (CD), and the newly developed middle frequency inverter DC. Usually the root mean square (RMS) values of the welding current are used in the machine parameter settings and the process controls. It is often the tedious job of the welding engineers to find the optimized welding current and time for each individual welding application.

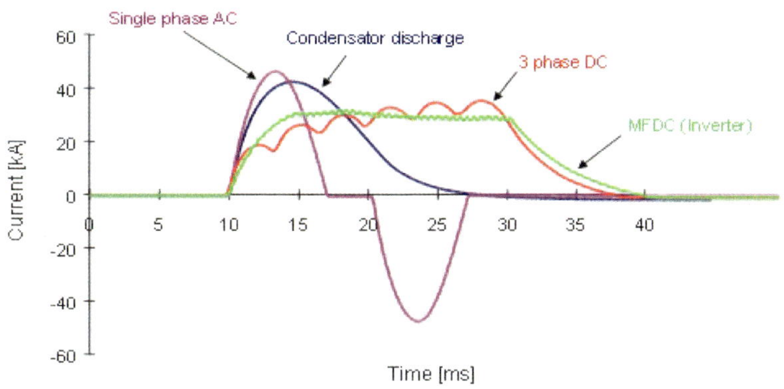

## 2) Welding Time

The heat generation is directly proportional to the welding time. Due to the heat transfer from the weld zone to the base metals and to the electrodes, as well as the heat loss from the free surfaces to the surroundings, a minimum welding current as well as a minimum welding time will be needed to make a weld. If the welding current is too low, simply increasing the welding time alone will not produce a weld. When the welding current is high enough, the size of the weld nugget increases with increasing welding time until it reaches a size similar to the electrode tip contact area. If the welding time is prolonged, expulsion will occur or in the worst cases the electrode may stick to the workpiece.

## 3) Welding Force

The welding force influences the resistance welding process by its effect on the contact resistance at the interfaces and on the contact area due to deformation of materials. The workpieces must be compressed with a certain force at the weld zone to enable the passage of the current. If the welding force is too low, expulsion may occur immediately after starting the welding current due to fact that the contact resistance is too high, resulting in rapid heat generation. If the welding force is high, the contact area will be large resulting in low current density and low contact resistance that will reduce heat generation and the size of weld nugget. In projection welding, the welding force causes the collapse of the projection in the workpiece, which changes the contact area and thereby the contact resistance and the current density. It further influences the heat development and the welding results.

## 4) Contact Resistance

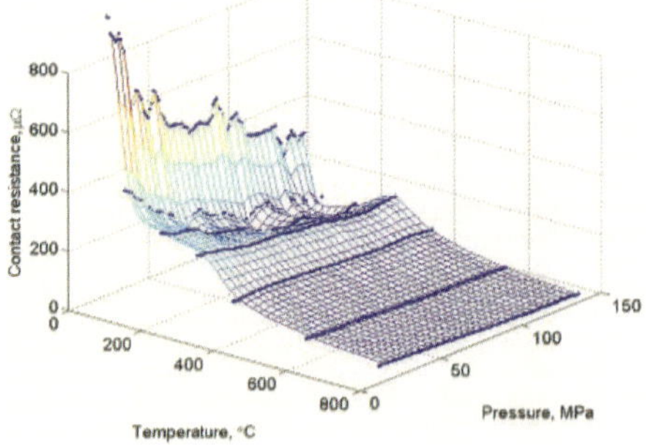


The contact resistance at the weld interface is the most influential parameter related to materials. It however has highly dynamic interaction with the process parameters. The figure below shows the measured contact resistance of mild steel at different temperatures and different pressures. It is noticed that the contact resistance generally decreases with increasing temperature but has a local ridge around 300°C, and it decreases almost proportionally with increasing pressure. All metals have rough surfaces in micro scale. When the welding force increases, the contact pressure increases thereby the real contact area at the interface increases due to deformation of the rough surface asperities. Therefore the contact resistance at the interface decreases which reduces the heat generation and the size of weld nugget. On the metal surfaces, there are also oxides, water vapour, oil, dirt and other contaminants. When the temperature increases, some of the surface contaminants (mainly water and oil based ones) will be burned off in the first couple of cycles, and the metals will also be softened at high temperatures. Thus the contact resistance generally decreases with increasing temperature. Even though the contact resistance has most significant influence only in the first couple of cycles, it has a decisive influence on the heat distribution due to the initial heat generation and distribution.

## 5) Materials Properties

Nearly all material properties change with temperature which add to the dynamics of the resistance welding process. The resistivity of material influences the heat generation. The thermal conductivity and the heat capacity influence the heat transfer. In metals such as silver and copper with low resistivity and high thermal conductivity, little heat is generated even with high welding current and also quickly transferred away. They are rather difficult to weld with resistance welding. On the other hand, they can be good materials for electrodes. When dissimilar metals are welded, more heat will be generated in the metal with higher resistivity. This should be considered when designing the weld parts in projection welding and selecting the forms of the electrodes in spot welding. Hardness of material also influences the contact resistance. Harder metals (with higher yield stress) will result in higher contact resistance at the same welding force due to the rough surface asperities being more difficult to deform, resulting in a smaller real contact area. Electrode materials have also been used to influence the heat balance in resistance welding, especially for joining light and non-ferrous metals.

## 6) Surface Coatings

Most surface coatings are applied for protection of corrosion or as a substrate for further surface treatment. These surface coatings often complicate the welding process. Special process parameter adjustments have to be made according to individual types of the surface coatings. Some surface coatings are introduced for facilitating the welding of difficult material combinations. These surface coatings are strategically selected to bring the heat balance to the weld interface. Most of the surface coatings will be squeezed out during welding, some will remain at the weld interface as a braze metal.

## 7) Geometry and Dimensions

The geometry and dimensions of the electrodes and workpieces are very important, since they influence the current density distribution and thus the results of resistance welding. The geometry of electrodes in spot welding controls the current density and the resulting size of the weld nugget. Different thicknesses of metal sheets need different welding currents and other process parameter settings. The design of the local projection geometry of the workpieces is critical in projection welding, which should be considered together with the material properties especially when joining dissimilar metals. In principle, the embossment or projection should be placed on the material with the lower resistivity in order to get a better heat balance at the weld interface.

## 8) Welding Machine Characteristics

The electrical and mechanical characteristics of the welding machine have a significant influence on resistance welding processes. The electrical characteristics include the dynamic reaction time of welding current and the magnetic / inductive losses due to the size of the welding window and the amount of magnetic materials in the throat. The up-slope time of a welding machine can be very critical in micro resistance welding as the total welding time is often extremely short. The magnetic loss in spot welding is one of the important factors to consider in process controls. The mechanical characteristics include the speed and acceleration of the electrode follow-up as well as the stiffness of the loading frame/arms. If the follow-up of the electrode is too slow, expulsion may easily occur in projection welding. The figure below shows measured process parameters in a projection welding process, which include the dynamic curves of the welding current, the welding force and the displacement of the electrode, where the sharp movement corresponds to the collapse of the projection in the workpiece.

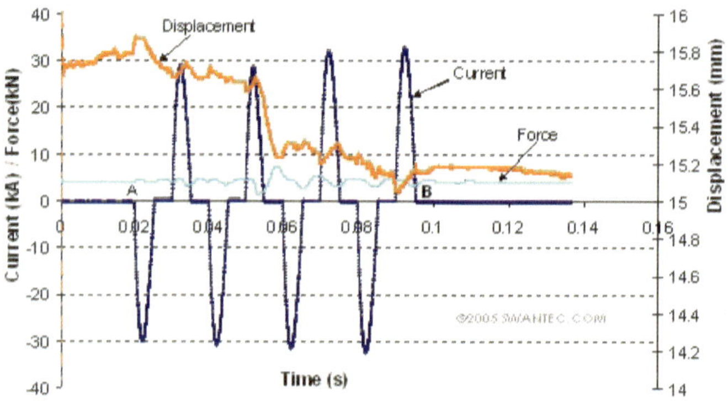


## Electrode Degradation and Tip Dressing

The resistance welding process is characterized with a high current passing through the materials to be welded between the electrodes under pressure for generating concentrated heat to form a weld. This highly concentrated heat also causes problems to the electrode tips with increasing number of welds.

### 1. Mechanisms of Electrode Degradation

The severe conditions of high current and pressure during resistance welding expose the electrode tips at a high risk of degradation. The photo to the right shows a comparison of the new and used electrode tips in spot welding of galvanized steel sheets. With increasing number of welds, there will be two major changes in the electrode tips:

- Geometric changes: The electrode tip diameter will increase due to deformation and wear, such as mushrooming, pitting or local material removal by picking up.
- Metallurgical changes: The material properties near the tip surface will also change during resistance welding such as alloying with sheet and coating materials, and recrystallization and softening by overheating.

New tip                Used tip

## 2. Effects of Electrode Degradation

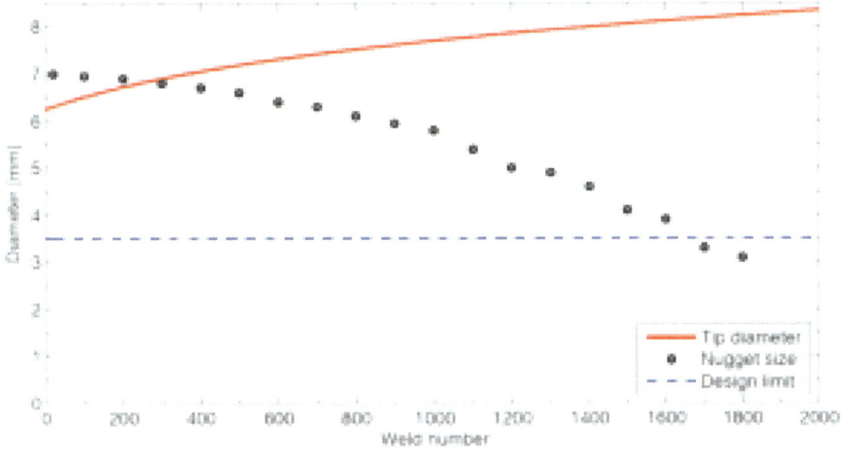


The increasing tip diameter will result in larger contact area between electrode and sheet thereby reducing the current density passing through the weld interface. At the same time, alloying of the electrode material with sheet and coating materials at the tip surface will reduce the conductivity of the electrode tip thereby also drag heat concentration away from the weld interface. Both effects lead to progressively reducing weld nugget sizes. After a certain number of welds, the resulted weld nugget will drop to below the minimum nugget size required for the weld quality as shown in the graph to the right. The number of welds achievable until the resulted weld nugget sizes dropping to the limit of weld quality is called the “electrode life”. This is dependent on the form and material of electrodes, the materials to be welded, surface coatings, and the interactions of dynamic welding process parameters.

## 3. Step Current and Tip-Dressing

Apart from adopting new materials and new designs of electrodes, two methods are generally used in production for compensating the electrode degradation in order to maintain the weld quality and increase the electrode life:

- Step current
- Electrode tip-dressing

Step current is a method to plan the spot welding process with a stepwise increasing weld current at each certain number of welds to compensate the loss of current density due to increasing tip diameter as shown in the graph to the right. The higher current needed with the larger tip diameter may be optimized through welding tests or by support of numerical simulations. In this way, more welds can be achieved without replacing electrodes hence a prolonged electrode life.

Electrode tip-dressing is a method to mechanically re-shape or abrasively clean the electrode tip after a certain number of welds to preserve nearly the same initial tip diameter and surface conditions. In this way, the welding process can be controlled at the same process parameters or slightly regulated by an adaptive control system to maintain consistent weld quality.

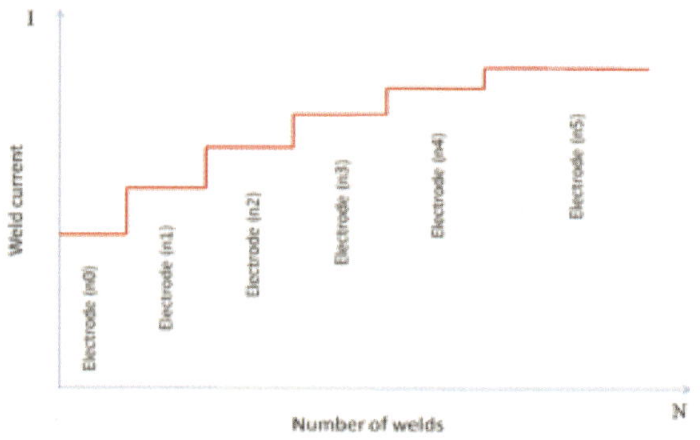


## 4. New Developments on tip-dressing and Optimization

SmartDress is a new tip-dressing system developed by the European consortium of seven partners including SWANTEC in the EC financed research project on: "Adaptive Tip Dress Control for Automated Resistance Spot Welding".

It will be a fully automated adaptive control system that will optimize, monitor, maintain electrode tip quality with combined mechanical cutting and abrasive cleaning, to maintain weld quality and minimize production line stoppages.

# Electric Resistance Welding

Electric resistance welding (ERW) produces coalescence of faying surfaces. Heat to form the weld comes from electrical resistance of material, rather than time and force. Factors influencing weld temperatures include work piece proportions, coating, electrode (material, geometry, pressing force), weld current, and weld time. Small pools are formed at the point of most resistance, the connecting surfaces. As a high current passes through metal, electrical resistance results.

## Spot Welding

Resistance welding joins overlapping metal sheets, studs, projections, electrical wiring hangers, some heat exchanger fins, or some tubing. Power sources and welding equipment should be sized to material and material thickness. Two copper electrodes clamp the sheets together and a current passes through. Heat is generated at the point of highest electrical resistance, where the surfaces touch. The rising temperature causes increased resistance and a molten metal pool between electrodes. When the current stops the metal solidifies under pressure. The process is accelerated by water cooled copper electrodes that remove surface heat.

## Seam Welding

ERW seam welding produces a weld at the faying surface. Two similar metals may be seamed at a butt joint or an overlap joint. Seam welding is often an automated process. Butt welding is similar, but differs in that butt welding welds the entire joint at once. Seam welding starts at one end and forms a weld progressively.

Seam welding relies on two electrodes. Usually made from copper, these electrodes provide the pressure or resistance as well as the current. The disc shaped electrodes rotate as the material passes between them. There is constant contact with the material for long continuous welds. In some cases the electrodes help move the material.

The seam is created when the electrical resistance of the work piece is higher than the rest of the circuit. The joint reaches melting point by the current. Semi-molten surfaces are pressed together. The welding pressure creates a fusion bond.

## Power Sources

Power sources include a transformer that supplies energy. The weld joint may be infused with low voltage or high current AC power.

## Cost

Equipment cost can be high. Cost per weld can be as little as four cents. Depending on application and manufacture rate, a return on your investment may still be seen in short order. Spot welding is also more cost effective than riveting.

## Advantages

ERW results in efficient energy use, limited work piece inconsistencies, high productivity, easy automation, little air pollution, and no filler requirements. A properly welded seam is often stronger than the original material. Durability is a result of a joint forged from heat and pressure.

## Limitations

Thickness is limited by welding power source output. If excessive heat is applied or heat is applied too quickly expulsion may occur. If the force between base materials is too low, the coating is too thick or too conductive, the high pressure will escape the containment force of the tips. Thinner metal will have less strength than welds without expulsion. ERW solid state bonds may result in joints with poor peel strength. Reflow braze bonds joining two dissimilar materials may result in bonds with poor peel and shear strength. The reflow braze process also requires a longer heating time.

## References


- Welding, technology: britannica.com, Retrieved 25 May 2018
- Different-types-of-arc-welding, welding: weldingschool.com, Retrieved 15 March 2018
- Arc-welding-processes-equipment: me-mechanicalengineering.com, Retrieved 25 June 2018
- What-is-arc-welding: wballoys.co.uk, Retrieved 15 April 2018
- The-fundamentals-of-gas-tungsten-arc-welding--preparation-consumables-and-equipment-necessary-for-the-process: thefabricator.com, Retrieved 22 March 2018
- Gas-metal-arc-welding, materials-science: sciencedirect.com, Retrieved 11June 2018
- An-overview-of-electric-resistance-welding: bakersgas.com, Retrieved 19 July 2018

# Permissions

All chapters in this book are published with permission under the Creative Commons Attribution Share Alike License or equivalent. Every chapter published in this book has been scrutinized by our experts. Their significance has been extensively debated. The topics covered herein carry significant information for a comprehensive understanding. They may even be implemented as practical applications or may be referred to as a beginning point for further studies.

We would like to thank the editorial team for lending their expertise to make the book truly unique. They have played a crucial role in the development of this book. Without their invaluable contributions this book wouldn't have been possible. They have made vital efforts to compile up to date information on the varied aspects of this subject to make this book a valuable addition to the collection of many professionals and students.

This book was conceptualized with the vision of imparting up-to-date and integrated information in this field. To ensure the same, a matchless editorial board was set up. Every individual on the board went through rigorous rounds of assessment to prove their worth. After which they invested a large part of their time researching and compiling the most relevant data for our readers.

The editorial board has been involved in producing this book since its inception. They have spent rigorous hours researching and exploring the diverse topics which have resulted in the successful publishing of this book. They have passed on their knowledge of decades through this book. To expedite this challenging task, the publisher supported the team at every step. A small team of assistant editors was also appointed to further simplify the editing procedure and attain best results for the readers.

Apart from the editorial board, the designing team has also invested a significant amount of their time in understanding the subject and creating the most relevant covers. They scrutinized every image to scout for the most suitable representation of the subject and create an appropriate cover for the book.

The publishing team has been an ardent support to the editorial, designing and production team. Their endless efforts to recruit the best for this project, has resulted in the accomplishment of this book. They are a veteran in the field of academics and their pool of knowledge is as vast as their experience in printing. Their expertise and guidance has proved useful at every step. Their uncompromising quality standards have made this book an exceptional effort. Their encouragement from time to time has been an inspiration for everyone.

The publisher and the editorial board hope that this book will prove to be a valuable piece of knowledge for students, practitioners and scholars across the globe.

# Index